TOPICS IN MODERN OPTICS

Chief Editor

A.S. PĀRASNIS

Co-editors

K.R. SARMĀ

R.K. THARÉJĀ

Formerly, Professors of Physics
Indian Institute of Technology
Kanpur

PUBLISHING FOR ONE WORLD

NEW AGE INTERNATIONAL (P) LIMITED, PUBLISHERS

New Delhi • Bangalore • Chennai • Cochin • Guwahati • Hyderabad
Jalandhar • Kolkata • Lucknow • Mumbai • Ranchi
Visit us at **www.newagepublishers.com**

Published by New Age International (P) Ltd., Publishers
First Edition: 2000
Reprint: 2009

Branches:

- 36, Malikarjuna Temple Street, Opp. ICWA, Basavanagudi, **Bangalore**. ✆ (080) 26677815
- 26, Damodaran Street, T. Nagar, **Chennai**. ✆ (044) 24353401
- CE-39/1016, Carrier Station Road, Near Paulson Park Hotel, Ernakulam South, **Cochin.** ✆ (0484) 4051303
- Hemsen Complex, Mohd. Shah Road, Paltan Bazar, Near Starline Hotel, **Guwahati**. ✆ (0361) 2543669
- No. 105, 1st Floor, Madhiray Kaveri Tower, 3-2-19, Azam Jahi Road, Nimboliadda, **Hyderabad**. ✆ (040) 24652456
- RDB Chambers (Formerly Lotus Cinema) 106A,1st Floor, S.N. Banerjee Road, **Kolkata**. ✆ (033) 22275247
- 16-A, Jopling Road, **Lucknow**. ✆ (0522) 2209578
- 142C, Victor House, Ground Floor, N.M. Joshi Marg, Lower Parel, **Mumbai**. ✆ (022) 24927869
- 22, Golden House, Daryaganj, **New Delhi**. ✆ (011) 23262370, 23262368

ISBN : 978-81-224-1242-0

Rs. 395.00

C-09-04-3649

Printed in India at Glorious Printers, Delhi.
Typeset at Jayanti Graphics Pvt. Ltd., New Delhi.

PUBLISHING FOR ONE WORLD
NEW AGE INTERNATIONAL (P) LIMITED, PUBLISHERS
4835/24, Ansari Road, Daryaganj, New Delhi-110002
Visit us at **www.newagepublishers.com**

PREFACE

Although OPTICS has been a course in the B.Sc. and M.Sc. curricula in Indian Universities for a century there has over the past three decades been a tendency to remove it altogether from M.Sc. curricula or reduce its weightage drastically in both BSc and MSc level curricula. At the same time the contents of the courses at all levels have remained the same—*viz* Huygens principle as an independent principle, double mirrors, Lloyd's mirror, biprism, Newton rings, single slit, calcite crystals, *etc*—as they were during the thirties and the forties, with nothing modified, nothing new added, no new methodology employed, no new attitudes displayed. The result is that the old and new optics and its fascinating applications which have emerged during the last twenty-five years have found no place in most curricula, indeed anywhere except in some of the IITs. The principles of optics clearly remain the same but, due to the advent of lasers, coherent optics, optical signal processing, communication optics, nonlinear optics, *etc*, their interpretation, usage and application have greater sophistication that is more conceptual than mathematical. The concepts are easily translatable at *all* levels—a point that has been insufficiently appreciated. Optics curricula in this country have largely remained untouched by the worldwide revolution in optics.[1]

Thus whatever optics is taught in India generally lacks both sophistication and motivation. In a university-level situation, unless Research and Development in optics is being done (as in only a few places), the numbers of students wanting to specialize in optics and cognate fields and Faculty who are in a position to teach the subject are small. Young graduates and scientists, even when they join laboratories and start working in optics and related fields, therefore, soon find themselves hopelessly handicapped by a lack of formal education. The only way open is self-study; however, there is always the possibility of there remaining large gaps in this preparation, even if genuine. There is also the natural tendency to pick what is immediately needed on the basis of exigency at the cost of a solid grounding of the fundamentals. What, therefore, is needed primarily is to make young scientists and engineers aware of the new optics and its applications in a way that would encourage motivated self-study.

---

1. See two recent papers by ASP: Teaching in modern ways: Optics, a case study–Parts I & II, 1997, *Physics Education (India)* **13**: 347-358 & **14**: 53-62.

There is in most academic and policy planning circles also a confusion about what optics really is. Some look upon it as an extension of a course in electrodynamics; others as a *service* discipline; still others as merely a kind of modern technology; yet others as something to do with light waves considered as some "disturbance". Many school and BSc curricula do not even mention that light waves are electromagnetic! The last three attitudes do not deserve recognition. The first is the most damaging since it prevails widely among people who should know better. It is necessary to dispel it by emphasizing the separate identity, character, spirit, methodology *and* technology of optics. For too long in India has there been a dichotomy between the science and technology of virtually everything, in this case of optics. Without technology optics would have only a soul but not a face just as without technology particle physics would not have the face of accelerators.

The proposal formulated in 1986 with this in mind by the Editors of this volume envisaged five Schools based respectively on *(a)* "Physical" optics, *(b)* Nonlinear and quantum optics, *(c)* Integrated optics and Fibre optical systems, *(d)* Interferometry and metrology, and *(e)* Teaching of optics. The purpose was to remedy the situation to some extent at the level of the working young scientist and university teacher. (Remedying the situation at the levels of school and college curricula is a more involved task that cannot be fulfilled by means of occasional intensive courses of a couple of weeks for just 30 participants.) The proposal was approved by the Science and Engineering Research Council of the Department of Science and Technology, Government of India, New Delhi, in mid-1986. The First School was held during 1986 November 10-22. It was believed that it would not only fill in the gaps and bring the young scientists to the required sophistication in frontier areas but also be instrumental in honing optical concepts and their usage. The theme was basic concepts of optics and optical processing. Since lasers are an integral part of anything in optics today the theme of optical resonators was included. Further, since nonlinear optical phenomena are playing a key and crucial role in optics in many different ways, they have also been included.

It is thirteen years since the first School was held. In 1990 July DST decided to support the publication of the lectures of SERC Schools. Various factors—for example, ASP's retirement from IIT Kanpur and delay at the hands of DST and the Publishers because of their own difficulties—have contributed to further delay in publishing these lectures. Since, however, the contents were, as said above, mostly based on *principles* it is thought that they would have direct relevance for all time to come. The lecturers were from among the most active workers in the field. All these contributions have been updated wherever essential. In some technologies discussed herein much progress has occurred in the last thirteen years and some parts of those topical chapters may be out of date. However, the basic technology as well as principles has remained the same. The contents have been organized as chapters in a natural sequence of development. Apart from the main and short courses there were colloquia by guest lecturers and also by some participants of the school; not being self-contained these are not included herein except Chapter 11.

This volume is offered to all young scientists, teachers and students who wish to be introduced to the fascinating field of modern optics with the hope that they would find the book at least partially useful.

A S PĀRASNIS
K R SARMĀ
R K THARÉJĀ

ACKNOWLEDGEMENTS

This School was sponsored generously by the Department of Science and Technology, Government of India, New Delhi, through its Science and Engineering Research Council. I I T Kanpur — in keeping with its tradition of a quarter of a century — supported the planning and execution of the School not only materially but also, and more importantly, by providing encouragement and freedom to the Centre for Laser Technology to propagate what may be called 'optics and laser culture'. In particular may be mentioned the assistance given by the then Director, Professor A K Mallik, the then Deputy Director, Professor G K Lal, the then Dean of Research and Development, Professor R N Biswas.

Lecturers planned their lectures in accordance with the philosophy[1] with which the School had been conceptualized. For the purposes of this Volume many of them unhesitatingly updated the contents as of 1996, notwithstanding the several years which had since passed.

When conducting the School help of several kinds was required and it was willingly provided by the PhD-students working with the Centre of Laser Technology: (now all *Dr*) Alikā Kharé, G Ravi Kumār, Vinay Kumār, G V Prasād, A Divākara Rāo, A Subbā Rāo, Sathaiah, Rékhā Tāmbé. The onerous task of organizing the diversified work of so many individuals was undertaken successfully by Dr Bansi Lāl, Senior Scientific Officer in the Centre, and A Subbā Rāo.

IIT Kanpur alumnus, and the creator of GIST (*Graphics Indian Scripts Terminal*) and its recent versions, Mohan Tāmbé, Centre for the Development of Advanced Computing, Government of India, Puné, now independent entrepreneur, Bangalore, helped with the software used for this book.

The staff of the Centre for Laser Technology led by Mr M H Nātu, Sr Superintendent, now Assistant Registrar, was always at hand to work and bail the organizers out of tricky situations.

All diagrams in this book—except those in Chapter 11—have been designed and drawn, based on the Author's original diagrams, by the Chief Editor, A S Pārasnis, who also edited all of the original lecture material for this book.

1. *See* Preface

Any work must be checked for errors: typing, uniformity of section and equation numbers, *references* at the end without reference to them in the text (and *vicé versâ*), occasional slips of language and syntax, and so on. Without the discerning, complete and prompt help given willingly by Professor D G Dhavalé (Retired, Fergusson College, Puné), notwithstanding the age of 88 years, who passed away in 1999 at the age of 93, this volume would have been the poorer. One of us (ASP) inherited a love for optics and learnt basic optics — not only the principles but also techniques like grinding, polishing, testing, figuring by employing the Foucault knife-edge test, chemical silvering, vacuum coating and aligning — from him during 1947-52. Then, as now, optical techniques were not part of students' education in optics. *He* learnt these in Allahabad in the late nineteen twenties when working with Méghnād Sāhā.

Last but not least thanks are due to the participants of the School who went through a rather rigorous exercise and contributed by their vigilance in asking questions which often made the lecturers think before answering. Some of them also gave *im promptu* short colloquia.

CONTENTS

LIST OF LECTURERS

MAIN COURSES

1 Professor A S Pārasnis
Department of Physics and
Head, Centre for Laser Technology
I I T Kanpur, KANPUR 208 016

2 Professor K R Sarmā
Dept of Electrical Engineering and
Centre for Laser Technology
I I T Kanpur, KANPUR 208 016

3 Professor R K Tharéjā
Department of Physics and
Centre for Laser Technology
I I T Kanpur, KANPUR 208 016

SHORT COURSES

4 Professor N C Māthur
Dept of Electrical Engineering
I I T Kanpur, KANPUR 208 016

5 Dr P C Méhtā
Instruments R & D Establishment
Raipur Road, DEHRADUN 248 008

6 Dr H C Pant
Laser and Plasma Physics Division
BARC, Trombay, BOMBAY 400 085

7 Dr K C Rustagi
Centre for Advanced Technology
Rajendra Nagar, INDORE 452 012

8 Professor B M Sivarām
Department of Physics
I I T Madras, MADRAS 600 036

COLLOQUIA

9 Dr B L Deekshātulu
Director
National Remote Sensing Agency
Balanagar, HYDERABDAD 500 037

10 Professor P R K Rao
Dept of Electrical Engineering
IIT Kanpur
[Notes on *Digital Signal Processing* not included]

11 Dr M U Siddiqui
Dept of Electrical Engineering
IIT Kanpur
[Notes on *Digital Signal Processing* not included]

ABSTRACTS

Chapter 1 Fourier Transforms

In this chapter are developed, without going into formal mathematics, a working knowledge and familiarity with Fourier transforms in order to enable the reader to follow the succeeding chapters which use them extensively. Concepts like convolution and correlations have also been developed. Several useful theorems have been listed and their use pointed out.

Chapter 2 Diffraction Theory

Classical scalar diffraction theory – which is the foundation of *superposition optics*[1]–is developed in a more or less customary but self-contained way, emphasizing physical concepts, and applied to screens with apertures of practical importance. The effect of finiteness of wavetrains has been derived for use in holography. Obstacles have not been considered because optical signal processing systems are usually composed of apertures in plane screens.

Chapter 3 Coherence Theory

Although coherence is one of the most important concepts in optics it is neglected in Physics curricula. It is usually dismissed by stating that "A sinusoidal wave is what a coherent wave is...". This "very correct" statement misses the modern point of view that the concept and its measure arise from the statistical characterization of real electromagnetic fields. This chapter develops the formalism of correlation and coherence functions and briefly points out its use. The close connection between coherence and polarization and between the corresponding matrices is pointed out at the end.

Chapter 4 Optical Signal Processing

The theory of one- and two-dimensional systems is developed for establishing system models for the optical elements: free space, lens and film transparency. The basic optical system, viz a lens receiving input signal and giving an output signal, is analyzed, thus deriving the imaging and Fourier transforming conditions. The operational method is used

---

1. a term coined by A S Parasnis as the title for his forthcoming textbook on Optics

to reanalyze it and to derive results in spatial filtering and complex mask-generating techniques. There is a section on incoherent signal processing.

Chapter 5 Holography

The first half is a non-mathematical account of holography. Recording, reconstruction and the "three-dimensional" character of holography are explained geometrically. The author's recent paper[1] is based on this part. This material can be understood by anyone who knows that light waves exist. The mathematical description of holography that follows is based on Chapter 2. The three-dimensional character of fringes in volume holograms and the operation of the Bragg equation have been clearly brought out. Diffraction efficiency and coupled wave theory have been touched upon.

Chapter 6 Holographic Interferometry

After describing how a combination of classical interferometry with holography leads to particularly useful techniques an account is given of elapsed time and real time methods. This is followed by a mathematical description of a variety of techniques: stored beam, stroboscopic stored beam, dynamic time average, double exposure, pulsed beam and other. The applications included here are: nondestructive testing, vibration studies, otological studies, bioholography in bats and dolphins.

Chapter 7 Holographic Optical Elements

The imaging properties of holograms[2] are increasingly receiving special attention and a new class of optical elements (HOE) is fast evolving. They can duplicate most of the functions provided by glass optics over a narrow spectral bandwidth. A review is given of the principles, design aspects, resolution, aberrations, unique characteristics of HOEs, HOE materials. The applications dealt with are: beam combiners, laser scanners, optical signal processing, and fibre optics and optical waveguide components.

Chapter 8 Holographic Scanners

For over a decade holographic scanners have been used in super-market universal product code readers. In recent times other applications have received much attention. After tracing the characteristics of different types of holograms and the advantages of holographic scanning this review goes on to give an account of optical considerations, parameter

1. Principles of Holography, Parasnis A S, 1990 *Physics Education* 7 181-193
2. The term ''imaging properties'' here does not refer to the reconstruction of the object wave by a read-out wave.

definitions, scan tasks, scanner types and holographic media. It ends by referring to manufacturing of practical systems.

Chapter 9 Radio Astronomical Interferometers

This chapter is a review of radio interferometers, with emphasis on the optical concepts and principles involved. After tracing the theory of interferometry, with special reference to the correlator interferometer, it deals in detail with aperture synthesis including supersynthesis, sampling, field of view and beam pattern and with applications like high resolution mapping and very large base interferometers (VLBI). The review ends by pointing out that in this field India is at the forefront and giving a brief account of the Ooty and Nārāyangāon interferometers.

Chapter 10 X-ray Imaging

This is a review of the principles and techniques of x-ray imaging. It begins with the pinhole camera, and discusses geometrical blurring and blur spot size in the short wavelength limit. The wave approximation and its diffractive effects are dealt with next and the two results compared to find a compromise. The designs of axisymmetric x-ray microscopes are then discussed. The last topics are zone plate-coded imaging and its limitations, and x-ray Fresnel lenses. It is pointed out that the key to achieving the x-ray capabilities of the Fresnel zone plate lies in the advancement of the technology for the fabrication of periodic microstructures with precision thickness and profile rather than new principles.

Chapter 11 Progress in Pattern Recognition and Image Processing

This account of recent progress in pattern recognition and image processing is based on the premise that the superiority of the human brain in pattern recognition is due not so much to the presence of some exotic pattern recognition operator as to its ability to decide which tools to use and in what combination. It deals with architectures, N-cubes, quad trees, problems and uses of texture and texture synthesis, segmentation and edge detection, artificial intelligence, and remote sensing.

Chapter 12 Optical Resonators

The concept of optical resonator and its detailed theory for the propagation of gaussian beams are developed to start with. Next the optical resonator is compared with lens waveguides. The stability of resonators is then dealt with in detail and criteria for stability given. The concept of modes is developed along with their connection with resonance frequencies. The emphasis throughout is on the optical principles involved and on the fact that lasers work on the basis of resonators.

Chapter 13 Nonlinear Optics

Nonlinear optics (NLO) today is too vast a field to be dealt with in a short article. This review traces certain aspects in a self-contained manner and points out the impetus received by NLO from the invention of the laser; there is hardly an area of Physics and technology that is untouched by NLO today. This chapter prepares the reader for the next chapter. Fundamentals arising out of the Maxwell equations and wave propagation in nonlinear media are dealt with rather than how the nonlinearity comes about. Second and third harmonic generation, optical mixing and frequency upconversion are also dealt with.

Chapter 14 Optical Phase Conjugation

A tutorial overview is presented of optical phase conjugation and degenerate four-wave mixing (DFWM). An outline of the physical phenomena involved is given first. This is followed by a brief introduction to the various applications of DFWM. Among the topics included are wavefront recovery and imaging applications of optical phase conjugation, the connection of DFWM with real time holography and transient gratings, and applications in optical computing. Diffusion-free nonlinearity in semiconductor-doped glasses is also discussed. To aid the reader some references from current literature have been included.

Chapter 1

FOURIER TRANSFORMS

A S PĀRASNIS

1.1 Introduction

Fourier Transforms are the common link between optics and communication theory. Whereas the information processed by the latter is usually temporal in nature that processed by optical methods is spatial. Apart from this, however, the similarity is so intimate that the formulation of both, based on Fourier analysis, is practically the same. Indeed the introduction of Fourier transforms in optics on a large scale—incidentally, by electrical engineers rather than physicists—has been instrumental in a good deal of recent development which is both scientific and technological. The *Physical Optics* of yesteryears is today's *Fourier* or *Coherent Optics* but with a complexion that is considerably different. Neither term is explanatory of the real nature of the phenomena encompassed. The first one emphasizes the mathematical rather than the physical aspects whereas the second one is unduly restrictive. At the same time it must be remembered that the basis of optics still remains the Maxwell equations and the principle of superposition. The present author has therefore coined a new term: *Superposition Optics*.

In this chapter our interest is to review the definitions, theorems and usage of Fourier transforms, convolutions and correlations, and thus gain a working familiarity with them. The proofs and existence theorems will be found in standard texts. A word about the notation: there are probably as many notations as authors! One has therefore to learn by sight to replace one set of symbols by another in integrals, formulæ, expressions, *etc*. For the imaginary unit $\sqrt{(-1)}$ physicists employ i whereas electrical engineers employ j. We choose the latter for the sake of uniformity and because increasing numbers of opticians are using it.

1.2 Definitions

The one-dimensional Fourier transform of $F(\xi)$, which may have real or complex values, of a real variable ξ that varies through infinite space of one dimension is defined as

$$\mathscr{F}\{F(\xi)\} \equiv f(x) = \int_{-\infty}^{\infty} F(\xi)\, e^{j2\pi x\xi}\, d\xi \qquad 1.1$$

About half of the authors put a negative sign in the exponent but this is of little consequence. The inverse operation gives

$$\mathscr{F}^{-1}\{f(x)\} = \mathscr{F}^{-1}\{\mathscr{F}\{F(\xi)\}\} = F(\xi)$$

$$= \int_{-\infty}^{\infty} f(x)\, e^{-j2\pi\xi x}\, dx \qquad 1.2$$

Although most optical functions—which represent the reality of electromagnetic fields—are well-behaved it is useful to keep in mind the following conditions for the existence of eq 1.1:

(a) F must be absolutely integrable over the infinite ξ-realm

(b) F must have only a finite number of discontinuities and a finite number of maxima and minima in any finite range of ξ

(c) F must have no infinite discontinuities

The two-dimensional Fourier transform of a function $F(\xi, \eta)$ of two variables is similarly defined, under similar conditions:

$$\mathscr{F}\{F(\xi,\eta)\} \equiv f(x,y) = \int_{-\infty}^{\infty}\int_{-\infty}^{\infty} e^{j2\pi(x\xi + y\eta)}\, d\xi\, d\eta \qquad 1.3$$

and the inverse transform as

$$\mathscr{F}^{-1}\{f(x,y)\} \equiv F(\xi,\eta)\} = \int_{-\infty}^{\infty}\int_{-\infty}^{\infty} f(x,y)\, e^{-j2\pi(\xi x + \eta y)}\, dx\, dy \qquad 1.4$$

$f(x)$ is called the *spectrum* of $F(\xi)$; similarly, $f(x, y)$ that of $F(\xi, \eta)$. Denoting $\sqrt{(\xi^2 + \eta^2)}$ by ρ and $\sqrt{(x^2 + y^2)}$ by r, eq 1.3 could be written as

$$\mathscr{F}\{F(\xi,\eta)\} = f(r) = \int_{0}^{\infty} F(\rho)\, e^{j2\pi r.\rho}\, dr \qquad 1.5$$

1.3 Some General Relationships

For quick calculations it is helpful to note the following general relationships which can easily be established:

Real space	FT space	
$F(\xi)$	$f(x)$	
$F(\xi, \eta)$	$f(x, y)$	
$F(-\xi)$	$f(-x)$	1.6
$F^*(\xi)$	$f^*(-x)$	1.7
$dF(\xi)/d\xi$	$(-j2\pi x)f(x)$	1.8
$d^n F(\xi)/d\xi^n$	$(-j2\pi x)^n f(x)$	1.9

1.4 Fourier Transform Theorems

(*i*) *Linearity Theorem*: The transform of a linear combination of F (ξ, η) and G (ξ, η) is the same combination of their individual transforms. Thus

$$\mathcal{F}\{\alpha F + \beta G\} = \alpha\mathcal{F}\{F\} + \beta\mathcal{F}\{G\} \qquad 1.10$$

(*ii*) *Similarity* (*or Reciprocity or Dilatation*) *Theorem*: If f(x,y) is the transform of F (ξ, η) we have

$$\mathcal{F}\{F(a\xi, b\eta)\} = \frac{1}{|ab|} f\left(\frac{x}{a}, \frac{y}{b}\right) \qquad 1.11$$

which, as an example, may be proved for one dimension:

$$\mathcal{F}\{f(a\xi)\} = \int_{-\infty}^{\infty} f(a\xi)\, e^{j2\pi x\xi}\, d\xi = \frac{1}{a}\int_{-\infty}^{\infty} F(X)\, e^{j2\pi\frac{x}{a}X}\, dX$$

which is nothing but (*1/a*) *f*(*x/a*). That is, a stretching/contraction of coordinates in real space leads to a contraction/stretching of coordinates in Fourier space. This fact alone makes us suspect that diffraction could be described by *Fourier transforms*.

(*iii*) *Shift* (*or Translation*) *Theorem*: If $F(\xi, \eta)$ is translated by displacements ξ_0, η_0 its transform is simply multiplied by exp $j2\pi(x\xi_0 + y\eta_0)$. In other words,

$$\mathcal{F}\{F(\xi - \xi_0, \eta - \eta_0)\} = e^{j2\pi(x\xi_0 + y\eta_0)}\,\mathcal{F}\{F(\xi, \eta)\}$$

for we have

$$\mathscr{F}\{F(\xi-\xi_0,\eta-\eta_0)\} = \int_{-\infty}^{\infty}\int_{-\infty}^{\infty} F(\xi-\xi_0,\eta-\eta_0)\, e^{j2\pi(x\xi+y\eta)}\, d\xi\, d\eta$$

$$\equiv \int_{-\infty}^{\infty}\int_{-\infty}^{\infty} F(X,Y)\, e^{j2\pi[x(X+\xi_0)+y(Y+\eta_0)]}\, dX\, dY$$

$$= e^{j2\pi(x\xi_0+y\eta_0)} \int_{-\infty}^{\infty} F(X,Y)\, e^{j2\pi(xX+yY)}\, dX\, dY$$

$$= e^{j2\pi(x\xi_0+y\eta_0)} \times \mathscr{F}\{F(\xi,\eta)\} \qquad 1.12$$

(*iv*) *Parseval's Theorem*: If $f(x,y)$ is the *FT* of $F(\xi,\eta)$ then

$$\int_{-\infty}^{\infty}\int_{-\infty}^{\infty} |F(\xi,\eta)|^2\, d\xi\, d\eta = \int_{-\infty}^{\infty}\int_{-\infty}^{\infty} |f(x,y)|^2\, dx\, dy \qquad 1.13$$

which in diffraction problems implies the conservation of energy as will be seen in due course. We shall note some more theorems a little later, after studying *convolutions*.

1.5 Convolutions

"Convolve" means "roll, fold or twist together". The *convolution* of two functions $F_1(\xi)$ and $F_2(\xi)$ is defined as

$$F_1(\xi) \otimes F_2(\xi) = \int_{-\infty}^{\infty} F_1(\xi')\, F_2(\xi-\xi')\, d\xi' \equiv F_2(\xi) \otimes F_1(\xi) \qquad 1.14$$

In words, $F_1(\xi')$ is multiplied by $F_2(\xi')$ after shifting the latter to an origin $\xi' = \xi$ and inverting to give $F_2(\xi-\xi')$; then the value of the product is integrated over ξ' and the result plotted as a function of ξ. To make the principle of the operation clear let us take a practical example. Suppose the intensity of a spectral line is measured by scanning it with a detector having a finite slit of width a as input aperture (Figure 1.1, next page). The coordinate ξ' may be the angle of scattering by a prism or grating and $I(\xi')$ shows the spectrum. The transmission function of the slit of width a is given by

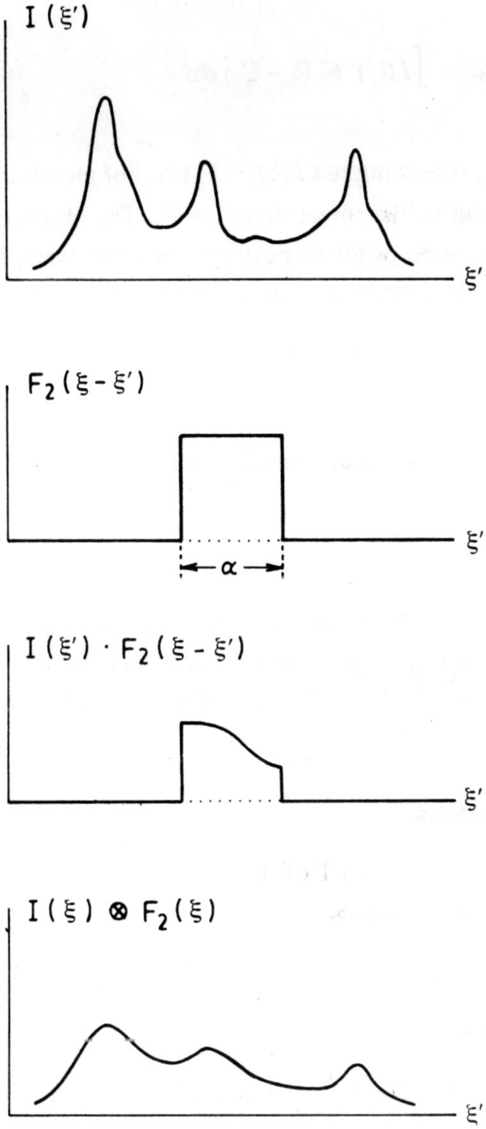

Figure 1.1 Concept of convolution. Curves represent, top to bottom, true spectrum, slit function, product for fixed value of variable, observed spectrum, respectively.

$$F_2(\xi') = 0\,, \ |\,\xi'\,| \geq a/2$$

$$= 1\,, \ |\,\xi'\,| < a/2$$

The slit thus transmits all radiation within a range a of ξ' and excludes all other. When it is placed at $\xi' = \xi$ the transmitted intensity as a function of ξ' is $I(\xi') \times F_2(\xi - \xi')$. What is recorded is the total transmitted intensity

$$I_{obs} = \int_{-\infty}^{\infty} I(\xi')\; F_2(\xi - \xi')\, d\xi'$$

which, plotted against ξ, is nothing but $I(\xi) \otimes F_2(\xi)$. For an arbitrary $I(\xi')$ as shown each part $d\xi'$ will be spread out by the spread function F_2. The observed intensity will lack the sharpness and resolution of the original spectrum. Another example is the spreading of the ideal image of a point object due to imperfect lenses:

$$I_{obs} = \delta(\xi, \eta) \otimes F_2(\xi, \eta) = F_2(\xi, \eta)$$

rather than $I_{obs} = \delta(\xi, \eta)$. (Here ξ, η are the x,y-coordinates of the image space.) For a finite object having an ideal image described by $I_{ideal}(\xi, \eta)$ the observed image would be given by

$$I_{obs} = I_{ideal}(\xi, \eta) \otimes F_2(\xi, \eta)$$

This also illustrates the spread function of an optical system as the response to a delta or *impulse* function. This provides the basis for the Green function method of diffraction theory. Yet another example is the Huygens principle itself expressed in the Kirchhoff formulation, as we shall see in the next chapter.

1.6 Further FT Theorems

(*v*) *Multiplication Theorem*: The FT of the product of two functions is the convolution of their individual transforms. That is,

$$\mathcal{F}\{F_1(\xi) \times F_2(\xi)\} = f_1(x) \otimes f_2(x) \qquad 1.15$$

Similarly in two dimensions.

(*vi*) *Convolution Theorem*: The FT of a convolution is the product of the individual Fourier transforms or

$$\mathcal{F}\{F_1 \otimes F_2\} = \mathcal{F}\{F_1\} \times \mathcal{F}\{F_2\} = f_1 \times f_2 \qquad 1.16$$

which follows from eq 1.15 by taking the inverse transform of both sides.

(*vii*) *Autocorrelation Theorem*: The autocorrelation of a function $F(\xi)$ is defined as a convolution with its conjugate with the sign of the argument changed, namely as

$$\int_{-\infty}^{\infty} F(\xi')\, F^*(\xi - \xi')\, d\xi' = F(\xi) \otimes F^*(-\xi) \qquad 1.17$$

Consequently we have a special case of eq 1.16, namely

$$\mathscr{F}\{F(\xi)\otimes F^*(-\xi)\} = f(x)\, f^*(x) = |f(x)|^2 \qquad 1.18$$

(*viii*) *Crosscorrelation theorem*: The crosscorrelation of $F_1(\xi)$ and $F_2(\xi)$ is defined as a convolution between one function and the conjugate with the sign of ξ changed in the latter, that is as

$$\int_{-\infty}^{\infty} F_1(\xi')\, F_2^*(\xi-\xi')\, d\xi' = F_1(\xi) \otimes F_2^*(-\xi) \qquad 1.19$$

From eq 1.16 we have

$$\mathscr{F}\{F_1(\xi) \otimes F_2^*(-\xi)\} = f_1(x)\times f_2^*(x) \qquad 1.20$$

It is needless to say that all results take relatively simple forms if the functions F, F_1, F_2 are real odd or real even.

Bibliography

FRANÇON M, 1979 *Optical Image Formation and Processing*, Translated by Bernard M Jaffe (New York: Academic Press), Appendix A

GOODMAN J W, 1968 *Introduction to Fourier Optics* (New York: McGraw-Hill Book Company), Chapter 2

HECHT E and ZAJAC A, 1979 *Optics* (4th Imp) (New York: Academic Press), Chapter 11

LIPSON S G and LIPSON H, 1969 *Optical Physics* (Cambridge: University Press), Chapter 3

Chapter 2

DIFFRACTION THEORY

A S PĀRASNIS

2.1 Introduction

Diffraction is that non-rectilinear propagation of light (in general, of *any* waves) that cannot be related to the phenomena of reflection and refraction. Diffraction phenomena are not amenable to exact theory so various assumptions and approximations are required to be made not only to simplify matters but also to make any progress at all. The main assumption here will be that the luminous field is a scalar field. Almost all results of interest here will still be available. The conditions which must be imposed upon the values of the field and those of its derivatives at boundary surfaces often appear artificial and inconsistent. They nevertheless give surprisingly accurate results which are well verified by experiment. The programme here will be to develop the Fresnel-Kirchhoff formula for the field at an arbitrary point behind an illuminated aperture, derive the Fraunhofer arrangement as a special case—at the same time showing that it gives the Fourier transform of the transmittance function of the aperture—and calculate the patterns for the more important practical situations.

2.2 The Helmholtz Equation

At any point P in diffraction space the electric field is written as

$$E(P; t) = Re\,[U(P)\,e^{-j\omega t}] \qquad 2.1$$

where, denoting the spatial part of the phase by $\phi(P)$,

$$U(P) = \mathrm{U}(P)\,e^{-j\phi(P)}$$

$E(P, t)$ is, of course, a solution of the wave equation

$$\nabla^2 E - \frac{1}{c^2}\frac{\partial^2 E}{\partial t^2} = 0 \qquad 2.2$$

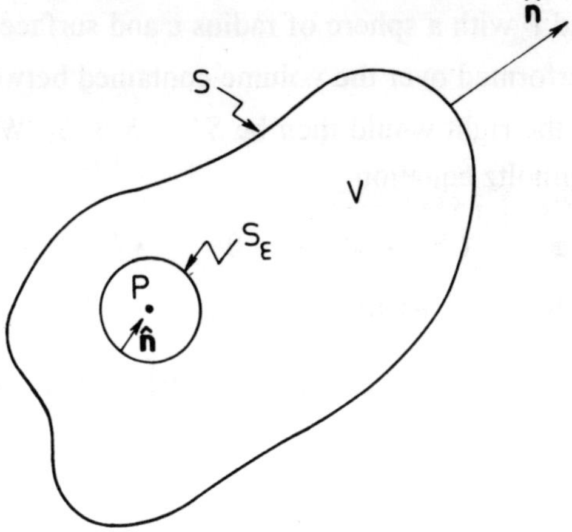

Figure 2.1 Volume V bounded by surface S around the field point P, and an infinitesimal sphere to exclude the singularity at the field point

Substituting eq 2.1 into eq 2.2 we get the *Helmholtz equation*

$$(\nabla^2 + k^2)\,U = 0 \qquad 2.3$$

$$k \equiv \left(\frac{\omega}{c}\right) = \left(\frac{2\,\pi}{\lambda}\right)$$

The usefulness of the Helmholtz equation lies in the assumption of monochromatic waves—that is, eq 2.1.

2.3 The Green Theorem

If $U(P)$, $G(P)$ be any two complex-valued functions of position and a surface S enclosing volume v surround the field point P (Figure 2.1) the Green theorem states that

$$\iiint_v \left(G\,\nabla^2 U - U\,\nabla^2 G\right) dv = \iint_S \left(G\,\frac{\partial U}{\partial n} - U\,\frac{\partial G}{\partial n}\right) dS \qquad 2.4$$

2.4 The Helmholtz-Kirchhoff Integral

The problem consists in finding the complex amplitude $U(P)$ at the field point when it is known at all points of S (Figure 2.1). Imagine there to be at P a source emitting spherical waves of amplitude at distance r given by

$$G = \frac{e^{-jkr}}{r} \qquad 2.5$$

To be able to use this as a Green function in eq 2.4 we must remove the discontinuity at P where $r = 0$. We surround P with a sphere of radius ε and surface S_ε. Then the integration at the left of eq 2.4 is performed over the volume contained between S and S_ε, say v'. The surface of integration at the right would then be $S' = S + S_\varepsilon$. Within v' G is a spherical wave that obeys the Helmholtz equation:

$$(\nabla^2 + k^2)\, G = 0 \tag{2.6}$$

Substituting eq 2.3,2.5 into eq 2.4 we get

$$\iiint \left(G\nabla^2 U - U\nabla^2 G \right) dv = - \iiint_{v'} \left(G\, U k^2 - U G k^2 \right) dv = 0 \tag{2.7}$$

andtherefore,

$$\iint_{S'} \left(G \frac{\partial U}{\partial n} - U \frac{\partial G}{\partial n} \right) dS = 0 \tag{2.8}$$

that is

$$\iint_{S} \left(G \frac{\partial U}{\partial n} - U \frac{\partial G}{\partial n} \right) dS = - \iint_{S_\varepsilon} \left(G \frac{\partial U}{\partial n} - U \frac{\partial G}{\partial n} \right) dS \tag{2.9}$$

For any point P_l on S' we have

$$G(r) = \frac{e^{-jkr}}{r}, \quad \frac{\partial G(r)}{\partial n} = - \cos(\boldsymbol{n}, \boldsymbol{r}) \left(jk + \frac{1}{r} \right) \frac{e^{-jkr}}{r} \tag{2.10a,b}$$

(**n**,**r**) being the angle between unit-**r** [≡ $P_0 P_1$] and the unit outward normal $\boldsymbol{n}$ at P_l. For P_l on S_ε, $(\boldsymbol{n}, \boldsymbol{r}) = \pi$ so $\cos(\boldsymbol{n}, \boldsymbol{r}) = -1$. Eq 2.10a,b then become

$$G = \frac{e^{-jk\varepsilon}}{\varepsilon}, \quad \frac{\partial G}{\partial n} = \left(\frac{1}{\varepsilon} + jk \right) \frac{e^{-jk\varepsilon}}{\varepsilon} \tag{2.11a,b}$$

The solid angle subtended by dS (which is at P_l) at P is $d\Omega = dS/\varepsilon^2$. The integral at the right of eq 2.9 then may be written as

$$\iint_{S_\varepsilon} \left(G \frac{\delta U}{\partial n} - U \frac{\partial G}{\partial n} \right) dS = \iint_{S_\varepsilon} \left[\frac{e^{-jk\varepsilon}}{\varepsilon} \frac{\partial U}{\partial n} - \frac{U e^{-jk\varepsilon}}{\varepsilon} \left(\frac{1}{\varepsilon} + jk \right) \right] \varepsilon^2 \, d\Omega \tag{2.12}$$

$$= \iint_{S_\varepsilon} - U \, d\Omega$$

$$= -4\pi U_P, \text{ as } \varepsilon \to 0 \tag{2.13}$$

Putting this into eq 2.9 and using the form of G chosen, we get

$$U_P = \frac{1}{4\pi} \iint_S \left\{ \frac{\partial U}{\partial n} \left(\frac{e^{-jkr}}{r} \right) - U \frac{\partial}{\partial n} \left(\frac{e^{-jkr}}{r} \right) \right\} dS \qquad 2.14$$

This expression for the field amplitude is called the *Helmholtz-Kirchhoff integral* the importance of which is that it allows the field at P to be evaluated directly in terms of the "boundary values" of the wave on any suitable closed surface surrounding the point.

2.5 Diffraction by a plane screen, Kirchhoff Conditions

The screen D in Figure 2.2 with an arbitrary aperture T is illuminated by a point source L. The problem is to evaluate eq 2.14 on the other side of the screen. Choose $S \equiv S_1 + S_2$,

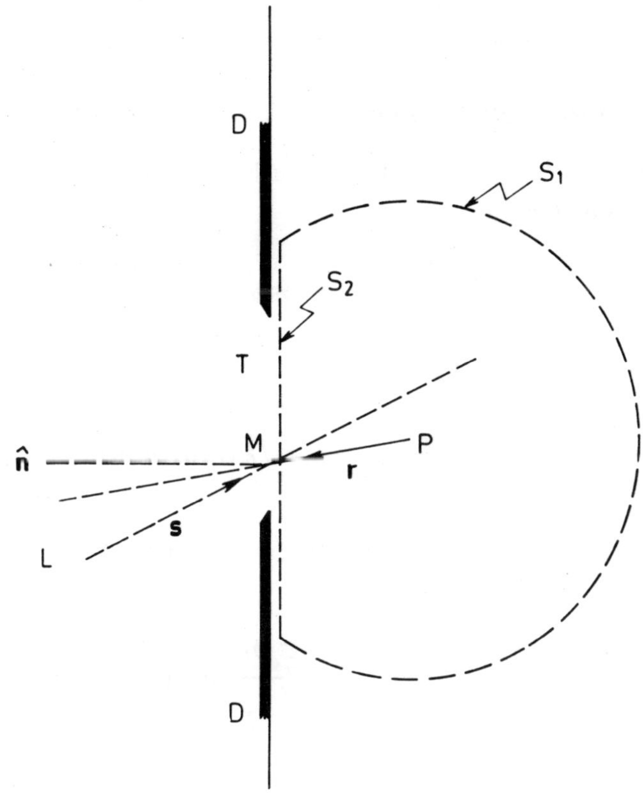

Figure 2.2 Diffraction by a plane aperture. Surface S made up of a large spherical cap and the part of the aperture plane intercepted by the sphere

S_1 being a sphere of radius R centred at P and S_2 the part of the screen intersected by it. In the integral of eq 2.14 put $S=S_1+S_2$ in the limits, **r** being PM. It is now necessary to know U and $\partial U/\partial n$ at all points of $S=S_1+S_2$. These are never known exactly so we have to make the following assumptions known as the *Kirchhoff boundary conditions*.

(*a*) U and $\partial U/\partial n$ are both taken to be zero on D

(*b*) U and $\partial U/\partial n$ have across T the same values as they would have in the absence of the screen

There remains the integration over S_1. Suppose the light source L begins to emit at time t_0; this would mean that it is not exactly monochromatic. By the time t the wave would have travelled a distance $c(t-t_0)$. Choose $R >> c\,(t-t_0)$ so that when the wave has reached P it has not reached S_1; therefore the contribution of S_1 to the integral is zero. Hence

$$U_P = \frac{1}{4\pi} \iint_{S_2 = T} \left\{ \frac{\partial U}{\partial n}\left(\frac{e^{-jkr}}{r}\right) - U \frac{\partial}{\partial n}\left(\frac{e^{-jkr}}{r}\right) \right\} dS \qquad 2.15$$

2.6 The Fresnel-Kirchhoff Formula

The distances r in the field are assumed to be much greater than the wavelength. Then we may approximate as follows:

$$\frac{\partial}{\partial n}\left(\frac{e^{-jkr}}{r}\right) \approx -\,jk\cos(\boldsymbol{n},\boldsymbol{r})\,\frac{e^{-jkr}}{r} \qquad 2.16$$

Substituting this into eq 2.15 we have

$$U_P = \frac{1}{4\pi} \iint_{T} \frac{e^{-jkr}}{r} \left[\frac{\partial U}{\partial n} + jkU\cos(\boldsymbol{n},\boldsymbol{r}) \right] dS \qquad 2.17$$

The wave from L has the form $\xi \exp(-jks)/s$, s being LM. Also, then, we have, with $U \equiv \xi \exp(-jks)/s$,

$$\frac{\partial U}{\partial n} = \frac{\partial}{\partial n}\left(\frac{\xi e^{-jks}}{s}\right) = -jk\frac{\xi e^{-jks}}{s} \cos(\boldsymbol{n},\boldsymbol{s})$$

Substituting into eq 2.17 we get the Fresnel-Kirchhoff formula:

$$U_P = \frac{jk\xi}{2\pi} \iint_{T} \frac{e^{-jk(r+s)}}{rs} \left[\frac{\cos(\boldsymbol{n},\boldsymbol{r}) - \cos(\boldsymbol{n},\boldsymbol{s})}{2} \right] dS \qquad 2.18$$

Writing

$$H(M) \equiv j\frac{k}{2\pi}\left(\frac{\xi e^{-jks}}{s}\right)[\cos(\boldsymbol{n}, \boldsymbol{r}) - \cos(\boldsymbol{n}, \boldsymbol{s})] \qquad 2.19$$

the formula of eq 2.18 for U_P becomes

$$U_P = \iint_T H(M)\,\frac{e^{-jkr}}{r}\,dS \qquad 2.20$$

The integrand may be interpreted as the wave, or *wavelet*, from the areal element at point *M* of the aperture T, *H(M)* being its amplitude at *M*, as it arrives at P a distance *r* away. The total field at P, that is the integral of eq 2.20, is the sum of such wavelets from points in T. This is the Huygens principle in essence, though not in detail. The differences between the quantitative, albeit approximate, calculation resulting in eq 2.19-20 and the original assertions of Huygens are as follows.

(*a*) *obliquity factor*: *H*(*M*) depends on not only *M* but also P, *i e* the direction of travel

(*b*) *phase shift*: The appearance of *j* in *H(M)* implies that there is a phase shift of $\pi/2$ between the source wave and the wavelet

(*c*) *wavelength-dependence*: The amplitude H(M) is inversely proportional to the wavelength. So, in the limit of short wavelength ($\lambda \rightarrow 0$), does U_P become infinite?

Eq 2.18 is symmetrical with respect to the field point P and the source L, so we get the *Helmholtz reciprocity theorem*: A source at L produces at P the same field as it would at L if placed at P. Further, *Babinet's principle for complementary screens* may be derived from eq 2.18. Complementary screens are a pair of apertures with opaque and transparent portions interchanged. Let $U_1(P)$, $U_2(P)$ be the fields produced by complementary screens, other things being identical. Let *U(P)* be the value when there is no aperture. Clearly, $U_1+U_2=U$. Hence if $U_1=0$ we have $U_2=U$, that is, at points where the irradiance $|U_1|^2$ is zero with one aperture it would be as if no aperture was there in the presence of the other. If $U=0$ we have $U_1 = -U_2$, *i e* the phases of U_1, U_2 differ by π and the irradiances are equal. Suppose a point source is imaged by an errorfree lens at L′; the light distribution will be zero in the image plane except in the immediate neighbourhood of L′. If complementary screens are placed between L and L′ we have $I_1=I_2$ except in the neighbourhood of L′.

2.7 Fraunhofer and Fresnel Diffraction

As *dS* explores over the aperture T, $\exp\{jk(r + s)\}$ in eq 2.18 will oscillate very rapidly since r+s will change by very many wavelengths but the obliquity factor will not vary much

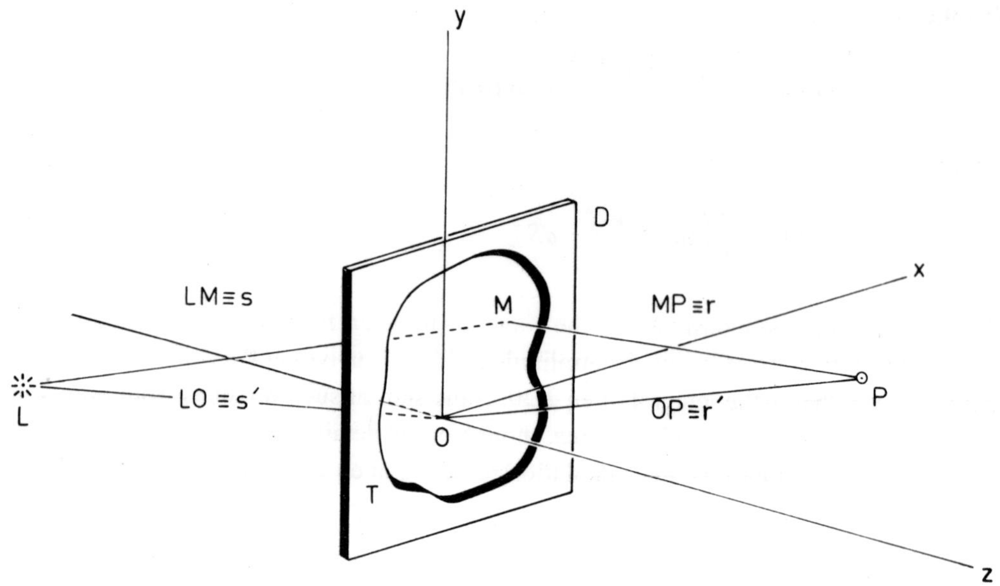

Figure 2.3 Diffraction of light from a point source

if $s, r, s', r' \gg |T|$ where the measure $|T|$ represents the dimensions of T (Figure 2.3). The angles (LO^LP) and (OP^LP) are small under the same conditions. Thus the angles (**n**^**r**) and (**n**^**s**) may be replaced by similar approximate values. The obliquity factor then will be $2\cos\delta$, δ being (LP^Oz). Finally, $1/rs$ may be replaced by $1/r's'$. Eq 2.18 then reduces to

$$U_P = \frac{jk\xi}{2\pi}\frac{\cos\delta}{r'\,s'}\iint_T e^{-jk(r+s)}\,dS \qquad 2.21$$

Let the coordinates of points be as follows:

$$L : X_0\,Y_0\,Z_0, \quad M: \xi\,\eta\,0, \quad P: x\,y\,z$$

Then we have

$$s^2 = (X_0 - \xi)^2 + (Y_0 - \eta)^2 + Z_0^2$$

$$s'^2 = X_0^2 + Y_0^2 + Z_0^2 \qquad 2.22\text{a,b}$$

and

$$r^2 = (x - \xi)^2 + (y - \eta)^2 + z^2 \qquad 2.23\text{a,b}$$

$$r'^2 = x^2 + y^2 + z^2$$

Or

$$s^2 = s'^2 - 2(X_0\xi + Y_0\eta) + \xi^2 + \eta^2$$

$$r^2 = r'^2 - 2(x\xi + y\eta) + \xi^2 + \eta^2 \qquad 2.24a,b$$

Expanding in powers of ξ/s', η/s', ξ/r', η/r', *etc.* we get

$$s = s' - \frac{X_0\xi + Y_0\eta}{s'} + \frac{\xi^2 + \eta^2}{2s'} - \frac{(X_0\xi + Y_0\eta)^2}{2s'^3} - \cdots \qquad 2.25a$$

$$r = r' - \frac{x\xi + y\eta}{r'} + \frac{\xi^2 + \eta^2}{2r'} - \frac{(x\xi + y\eta)^2}{2r'^3} - \cdots \qquad 2.25b$$

Substituting into eq 2.21 we have

$$U_P = \frac{jk\cos\delta}{2\pi}\,\frac{\xi\, e^{-jk(s'+r')}}{s'\,r'} \iint_T e^{-jkf(\xi,\eta)}\, d\xi\, d\eta \qquad 2.26$$

where, for the sake of brevity, we have defined $f(\xi, \eta)$ in the exponent as

$$f(\xi,\eta) \equiv -\frac{X_0\xi + Y_0\eta}{s'} - \frac{x\xi + y\eta}{r'} + \frac{\xi^2 + \eta^2}{2s'} + \frac{\xi^2 + \eta^2}{2r'}$$
$$- \frac{(X_0\xi + Y_0\eta)^2}{2s'^3} - \frac{(x\xi + y\eta)^2}{2r'^3} \cdots \qquad 2.27$$

Denote the direction cosines of L and P as

$$L : -\frac{X_0}{s'} \equiv l_0, \quad -\frac{Y_0}{s'} \equiv m_0; \quad P : \frac{x}{r'} \equiv l, \frac{y}{r'} \equiv m$$

Then eq 2.27 becomes

$$f(\xi,\eta) = (l_0 - l)\,\xi + (m_0 - m)\,\eta$$
$$+ \frac{1}{2}\left\{\left(\frac{1}{s'} + \frac{1}{r'}\right)(\xi^2 + \eta^2) - \frac{(l_0\xi + m_0\eta)^2}{s'} - \frac{(l\xi + m\eta)^2}{r'}\right\} \cdots \qquad 2.28$$

The evaluation of the integral of eq 2.26 is simpler when the quadratic and higher order terms are negligible. The function $f(\xi, \eta)$ then varies linearly across the aperture and we have *Fraunhofer diffraction*. When terms higher than the quadratic are negligible we have *Fresnel diffraction*. The unfortunate custom of calling anything that is not *Fraunhofer* diffraction *Fresnel* diffraction should be abandoned. The quantitative, hence correct, terminology is in the preceding statement, for in general we have situations which are not amenable to any kind of quantitative formulation. Later we shall relate these terms to the actual geometry.

The quadratic and higher terms disappear only in the limit $s', r' \to \infty$. However, the former do not contribute much to the integral if

$$\frac{1}{2}k\left|\left[\frac{1}{s'}+\frac{1}{r'}\right](\xi^2+\eta^2)-\frac{(l_0\xi+m_0\eta)^2}{s'}-\frac{(l\xi+m\eta)^2}{r'}\right| << 2\pi \qquad 2.29$$

Many situations follow from this. For example, eq 2.29 will be satisfied if

$$|r'|,\ |s'| >> \frac{(\xi^2+\eta^2)_{max}}{\lambda} \qquad 2.30$$

or

$$\frac{1}{s'}+\frac{1}{r'} = 0 \text{ and } l_0^2, m_0^2, l^2, m^2 << \frac{|r'|\lambda}{(\xi^2+\eta^2)_{max}} \qquad 2.31$$

These conditions are important when the aperture T is a lens. Eq 2.26 may under conditions of Fraunhofer diffraction be written as

$$U_P = C\iint_T T(\xi,\eta)\, e^{-jk\{(l_0-l)\xi+(m_0-m)\eta\}}\, d\xi\, d\eta$$

or, putting $p \equiv l - l_0$ and $q \equiv m - m_0$,

$$U_P = C\iint_T T(\xi,\eta)\, e^{jk(p\xi+q\eta)}\, d\xi\, d\eta \qquad 2.32$$

A factor $T(\xi,\eta)$ has been introduced here to indicate explicitly the *transmittance function* of the aperture. Its meaning is that if the incident amplitude at a region around a point ξ,η of the aperture is unity the transmitted one will be $T(\xi,\eta)$. $T(\xi,\eta)$ is complex in the event of the region being retardive as well as absorptive. From eq 2.32 we see that the Fraunhofer diffraction pattern of a plane aperture is the Fourier transform of its transmittance function and all of the theorems of Chapter 1 may be applied to all problems involving Fraunhofer diffraction (*see* Chapters 4-8).

2.8 Special Cases of Importance

In order to avoid a continual connotation that Fourier transforms in optics must always involve only distances we shall define the Fourier transform of $F(\xi, \eta)$ (*see* Chapter 1) in terms of non-committal symbols λ, μ as

$$f(\lambda, \mu) \equiv \mathscr{F}\{F(\xi, \eta)\} = \int_{-\infty}^{\infty}\int_{-\infty}^{\infty} F(\xi, \eta)\, e^{j2\pi(\lambda\xi + \mu\eta)}\, d\xi\, d\eta \qquad 2.33$$

On the other hand, the electric field amplitude in the diffraction field is

$$U_P = C \iint_T T(\xi, \eta)\, e^{jk(p\xi + q\eta)}\, d\xi\, d\eta \qquad (2.32)$$

Analogously with the temporal angular frequency (or *pulsation*) ω of $e^{j\omega t}$ the quantities kp and kq are called *spatial angular frequencies* (or *spatial pulsations*), and λ and μ—which would correspond with $\nu = \omega/2\pi$—simply *spatial frequencies*. In what follows we shall see what the physical significance of these quantities is. From the definitions of p, q and l_0, l, m_0, m we get

$$p \equiv l - l_0 = \frac{x}{r'} + \frac{X_0}{s'} \text{ and } q \equiv m - m_0 = \frac{y}{r'} + \frac{Y_0}{s'} \qquad 2.34$$

or, if the source L is at infinity, i.e. $s' \to \infty$, we have

$$p = \frac{x}{r'} \text{ and } q = \frac{y}{r'} \qquad 2.35$$

2.8.1 Rectangular Aperture (Figure 2.4, width *a*, height *b*)

Taking the origin of ξ, η at the centre of the opening we write the transmittance function as

$$T(\xi, \eta) = 1 \text{ for } |\xi| < \frac{a}{2}, |\eta| < \frac{b}{2} \qquad 2.36$$

$$= 0 \textit{ otherwise}$$

whereby the field amplitude at P becomes

$$U_P = \int_{-a/2}^{a/2}\int_{-b/2}^{b/2} e^{jk(p\xi + q\eta)}\, d\xi\, d\eta = ab\,\frac{\sin(kpa/2)}{kpa/2}\,\frac{\sin(kqb/2)}{kqb/2}$$

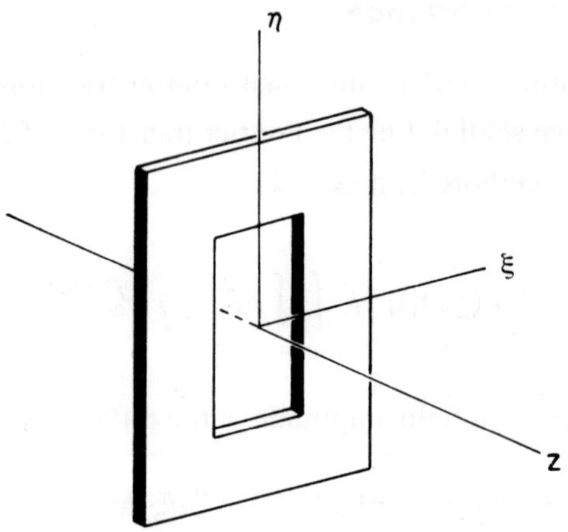

Figure 2.4 Diffraction by a rectangular aperture

or, denoting $(kpa/2) = (\pi\, a/\lambda)\,(x/r') = (\pi\, a/\lambda)\sin\theta_x$ by α and $(kqb/2) = (\pi\, b/\lambda)\sin\theta_y$ by β,

$$U_P = a\,b\,\frac{\sin\alpha}{\alpha}\,\frac{\sin\beta}{\beta} \tag{2.37}$$

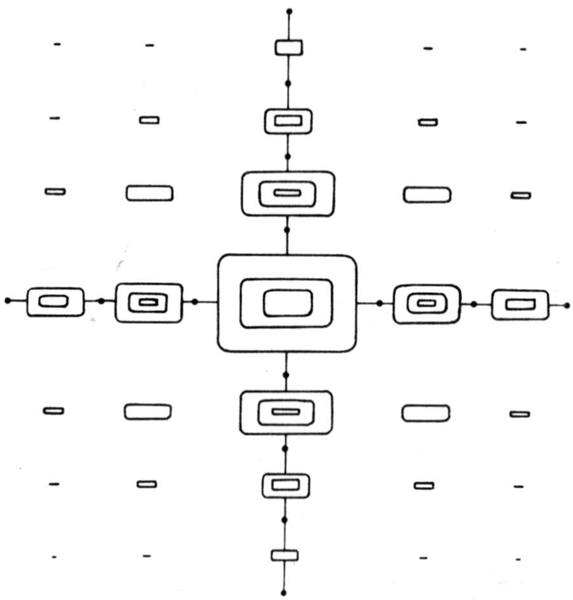

Figure 2.5 Fraunhofer diffraction pattern of a rectangular aperture

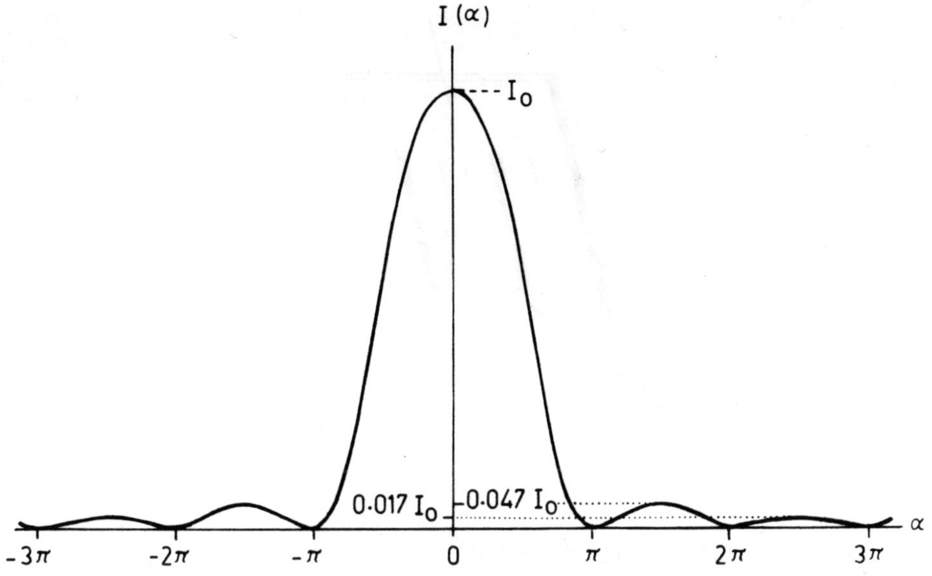

Figure 2.6 Diffraction pattern irradiance curve of a narrow slit

so that the distribution in the diffraction field is given by

$$I_P \equiv |U_P|^2 = I_0 \frac{\sin^2\alpha}{\alpha^2}\frac{\sin^2\beta}{\beta^2} \quad 2.38$$

I_0 being the maximum value, at $\alpha, \beta = 0$ *i.e.* along O*z*. Each diffraction maximum would give on a distant screen a patch of light. A visually useful representation is given in Figure 2.5 on p. 18. The well known *single slit* formula follows if we put $b \to 0$ for then $\beta \to 0$ and we get

$$I_P = I_0 \frac{\sin^2\alpha}{\alpha^2}, \; where \; \alpha \equiv \frac{\pi a \sin\theta}{\lambda} \quad 2.39$$

which is depicted in Figure 2.6. Clearly the minima occur at $a \sin\theta_x = (\text{integer}) \times \lambda$ and the secondary maxima are the solutions of $\tan\alpha = \alpha$ as may readily be found.

2.8.2 Two Slits (Figure 2.7, width *a*, distance *A* apart, *see next page*)

If the transmittance function of a single slit (*i. e.* eq 2.36 with η deleted) is denoted $s(\xi)$ we may write that of a double slit as

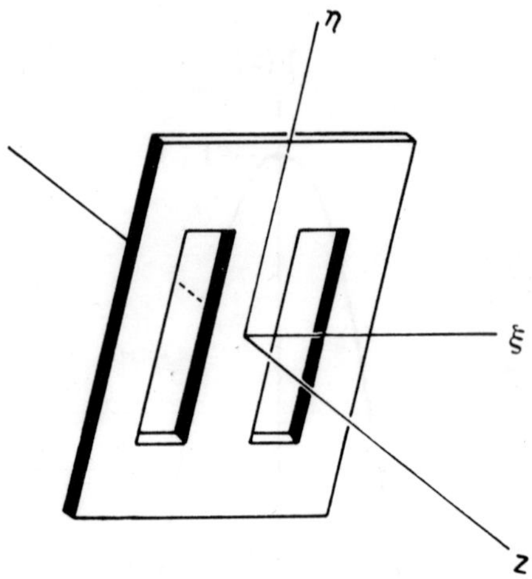

Figure 2.7 Two identical rectangular apertures

$$T(\xi) = s(\xi) \otimes \left[\delta\left(\xi + \frac{A}{2}\right) + \delta\left(\xi - \frac{A}{2}\right)\right] \tag{2.40}$$

The Fraunhofer diffraction amplitude is given by the Fourier transform as

$$U_P = \mathscr{F}\{T(\xi)\} = \mathscr{F}\{s(\xi)\} \,.\, \mathscr{F}\left\{\left[\delta\left(\xi + \frac{A}{2}\right) + \delta\left(\xi - \frac{A}{2}\right)\right]\right\}$$

which may be worked out as

$$= a\,\frac{\sin\alpha}{\alpha}\left(e^{-jkpA/2} + e^{jkpA/2}\right)$$

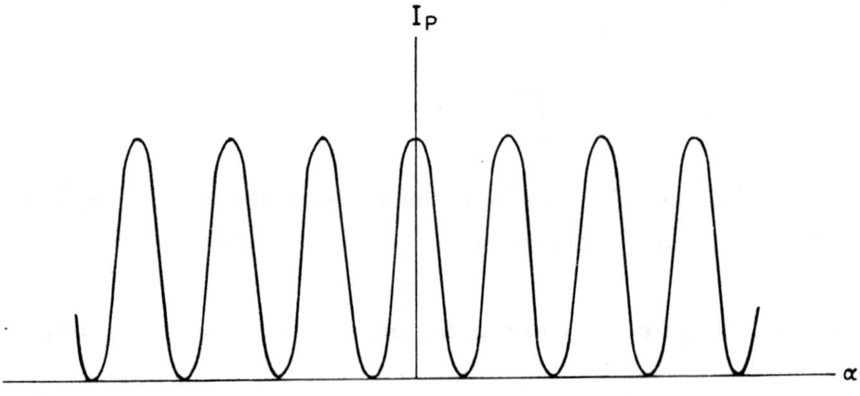

Figure 2.8 Diffraction pattern irradiance curve of two narrow slits

$$= 2a \frac{\sin\alpha}{\alpha} \cos(kpA/2)$$

$$\equiv 2a \frac{\sin\alpha}{\alpha} \cos\gamma \qquad 2.41$$

The irradiance pattern then is given by

$$I_P = I_0 \left(\frac{\sin^2\alpha}{\alpha^2}\right)(\cos^2\gamma) \qquad 2.42$$

To get the *two equal narrow slits formula, i. e.* Young fringe formula, put a=0:

$$I_P = I_0 \cos^2\gamma \qquad 2.43$$

—the well known *cosine-squared* distribution shown in Figure 2.8 on the last page.

2.8.3 Circular Aperture (Figure 2.9, diameter *a*)

The transmittance function is

$$T(\xi, \eta) = 1 \text{ for } (\xi^2 + \eta^2)^{1/2} < \frac{a}{2},$$

$$= 0 \textit{ otherwise}$$

Converting to plane polar coordinates u and θ we get

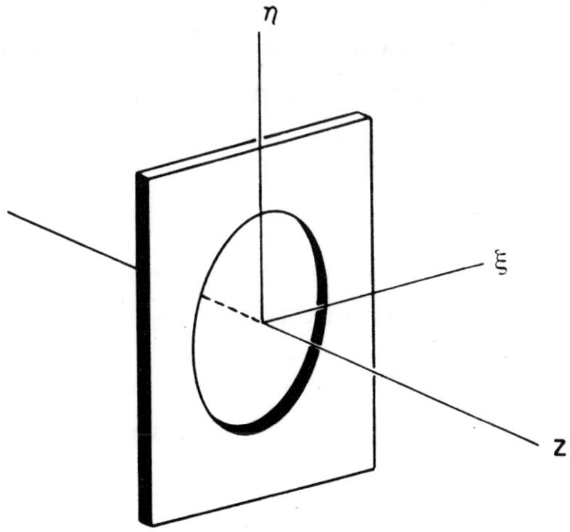

Figure 2.9 Circular aperture

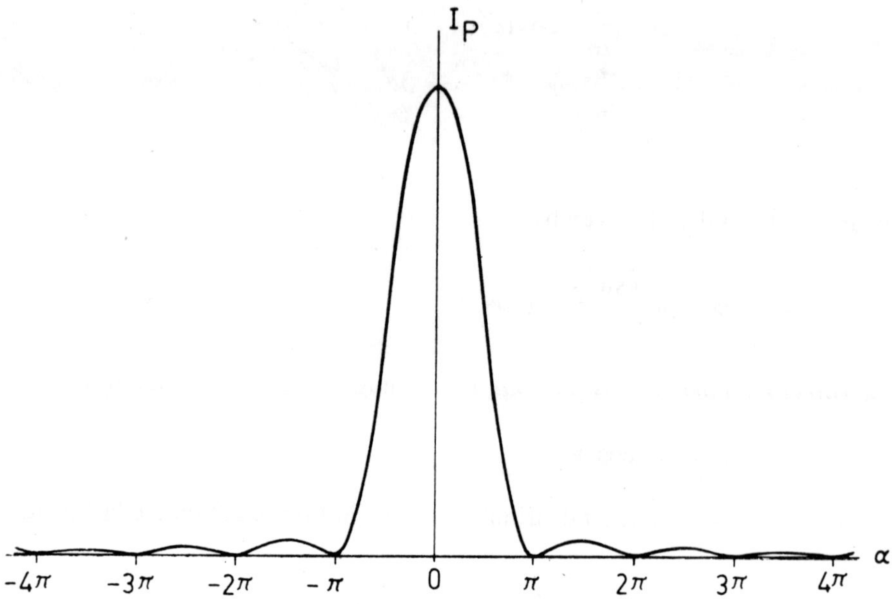

Figure 2.10 Diffraction pattern irradiance curve of a circular aperture

$$U_P = \left(\frac{\pi a^2}{2}\right) \frac{J_1(\pi a u)}{\pi a u} = \left(\frac{\pi a^2}{2}\right) \frac{J_1(k p a/2)}{k p a/2} \qquad 2.44$$

$$I_P = I_0 \left[\frac{J_1(k p a/2)}{k p a/2}\right]^2 \qquad 2.45$$

This socalled *Airy pattern* is shown in Figure 2.10. Its importance lies in the fact that circular lenses and other circular apertures constitute almost 90% of all apertures used in optics.

2.8.4 Finite Wave Train

2.8.4.1 Slit Aperture (*see* Figure 2.11, next page)

A slit aperture at $\xi = 0$ and parallel to the η-axis is described by the transmittance function $T(\xi) = \delta(\xi)$ for all η. If it were at $\xi = \xi_0$ we would have $T(\xi) = \delta(\xi - \xi_0)$. Then the diffraction amplitude would be given by

$$\mathscr{F}\{\delta(\xi)\} \equiv t(p) = 1$$

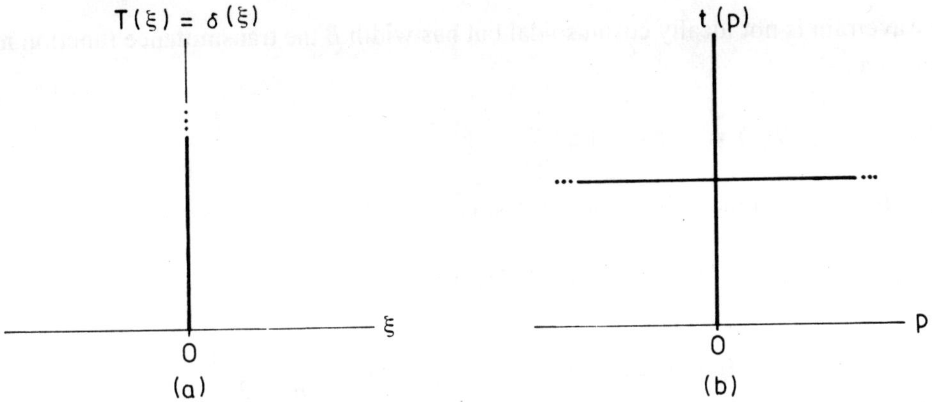

Figure 2.11 (a) Transmittance function T of an infinitesimal slit, which is impulse or delta function, (b) Fourier transform of the delta function

Conversely

$$\mathscr{F}\{1\} = \delta(p)$$

In general, the Fourier transform of a constant is a delta function. In calculating the diffraction amplitude of this slit aperture as being of uniform irradiance we have here ignored the obliquity factor that is contained in C of eq 2.32. Especially here it is not constant. However, even if it were to be considered the irradiance would diminish as $1/R$ with increasing $|p|$, the essential feature, *viz* the absence of maxima/minima, would remain the same. A point aperture at the origin is described by $T(\xi, \eta) = \delta(\xi, \eta)$ and that at ξ_0, η_0 by $T(\xi, \eta) = \delta(\xi - \xi_0, \eta - \eta_0)$. The Fraunhofer diffraction pattern then corresponds to the Fourier transform of $\delta(\xi, \eta)$, again a constant $t(p,q)=1$—apart from the $1/r$-variation. These results confirm what is seen more or less intuitively.

2.8.4.2 Two Narrow Slits (or two point apertures)

We get the Young pattern by putting a, hence α, equal to zero in eq 2.41, *i.e.* $U_p = \cos(k A p/2)$, not worrying about a constant in front. That is, if $T(\xi) = \delta(\xi + A/2) + \delta(\xi - A/2)$, taking the Fourier transform the diffraction pattern would correspond to $\cos(kAp/2)$. Conversely, if $T(\xi) = \cos(k A \xi/2)$ we have its Fourier transform as $\delta(p + A/2) + \delta(p - A/2)$. That is, if the transmission through an aperture is cosinusoidal the diffraction pattern is two narrow slit-like maxima.

2.8.4.3 Finite Wavetrain

If the wavetrain is not ideally cosinusoidal but has width B the transmittance function may be written as

$$T(\xi) = 2\cos(kA\xi/2).\, s_B(\xi) \tag{2.46}$$

where $s_B(\xi)$ is a slit function of width B. Using the multiplication theorem,

$$\mathscr{F}\{T(\xi)\} = \mathscr{F}\{2\cos(kA\xi/2)\} \otimes \mathscr{F}\{s_B(\xi)\}$$

$$= [\delta(p + A/2) + \delta(p - A/2)] \otimes B\,\frac{\sin(kBp/2)}{kBp/2} \tag{2.47}$$

so that the diffraction pattern is two slit patterns centred at $\pm A/2$ respectively. The irradiance distribution is given by

$$I_P = \left[B^2\,\frac{\sin^2(kBp/2)}{(kBp/2)^2} \right] \otimes [\delta(p + A/2) + \delta(p - A/2)] \tag{2.48}$$

Clearly this approximate formula is not valid if the length of the wavetrain is only a few wavelengths. Also note that if the wavetrain is asymmetric, *i. e.* that of Figure 2.12 is moved through (1/4 $1/A$), one of the *sinc*-functions of *t(p)* will be inverted; 'which?' depends on the displacement being to the left or the right.

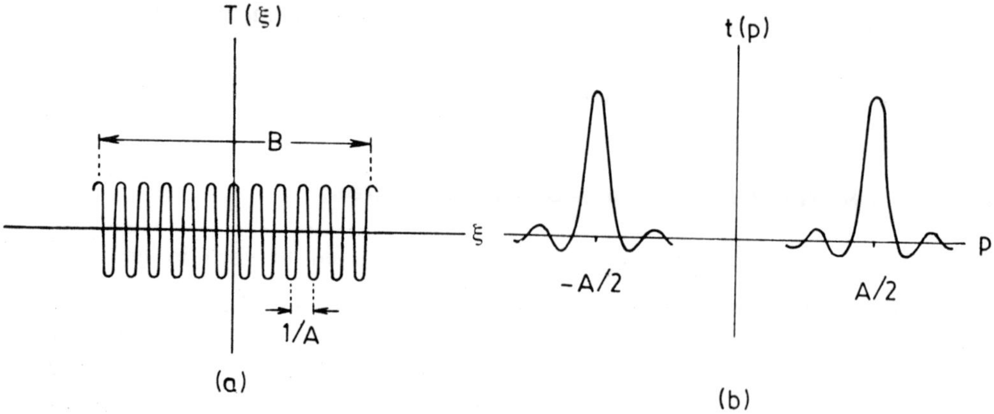

Figure 2.12 (a) Transmittance function that is a finite cosine function, (b) FT of the same

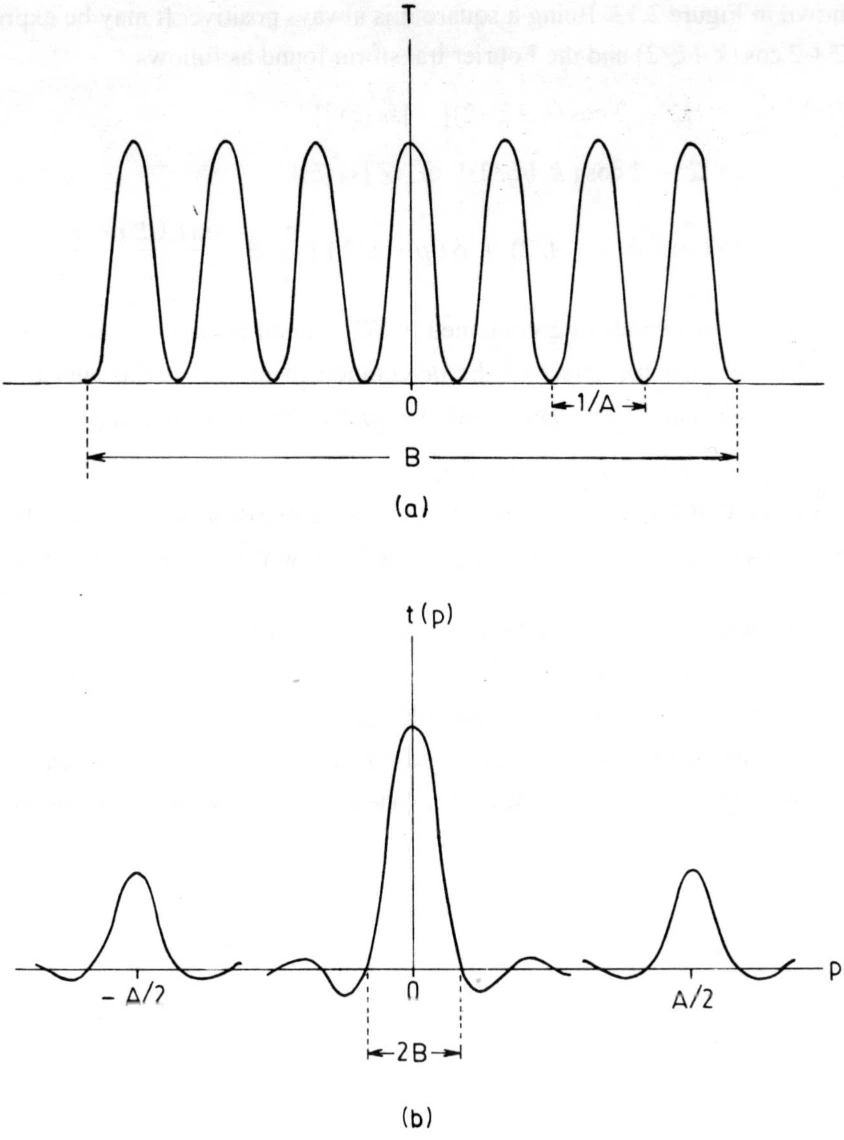

Figure 2.13 (a) Transmittance function that is a finite cosine function with a "dc" component, (b) Fourier Transform of the same

2.8.4.4 Finite Wavetrain with a dc Component

Suppose that we have a transmittance function along with a slit function of width B given by

$$T(\xi) = 4\cos^2(kA\,\xi/4)\,.\,s_B(\xi) \qquad 2.49$$

This is shown in Figure 2.13. Being a square it is always positive. It may be expressed as $T(\xi) = 2 + 2\cos(kA\xi/2)$ and the Fourier transform found as follows:

$$\begin{aligned}\mathscr{F}\{T(\xi)\} &= \mathscr{F}\{[2 + 2\cos(kA\xi/2)] \cdot [s_B(\xi)]\} \\ &= \mathscr{F}\{2 + 2\cos(kA\xi/2)\} \otimes \mathscr{F}\{s_B(\xi)\} \\ &= [\delta(p) + \delta(p + A/2) + \delta(p - A/2)] \otimes B\,\frac{\sin(kBp/2)}{kBp/2}\end{aligned} \qquad 2.50$$

Again, the number of periods of ξ contained in $T(\xi)$ should be sufficiently large (*see* Fig 2.13). We then see that the effect of the *dc*-component in $T(\xi)$ is the appearance of a *sinc*-function $\delta(p) \otimes sinc$ at $p=0$. This result has applications in *Holography*.

Bibliography

BORN M and WOLF E, 1975 *Principles of Optics* 5th ed (London: Pergamon Press), Chapter VIII

FRANÇON M, 1979 *Optical Image Formation and Processing* (New York: Academic Press), Appendix B, Chapter 3

GOODMAN J W, 1968 *Introduction to Fourier optics* (New York: McGraw-Hill), Chapters 3 & 4

HECHT E, 1987 *Optics*, 2nd ed (New York: Academic Press)

IIZUKA K, 1987 *Engineering Optics* (Berlin: Springer-Verlag), Chapters 3 & 4

KLEIN M V and FURTAK T E, 1986 *Optics*, 2nd ed (New York: Wiley-Interscience), Chapter 6

WOOD R W, 1988 *Physical Optics*, 3rd ed (Washington: Optical Society of America), Chapter VII

Chapter 3

COHERENCE THEORY

K R SARMA

3.1 Introduction[1]

The electric field at any point P with coordinates (x, y, z) is a time-varying vector and may be expressed as

$$\mathbf{E}(P, t) = \mathbf{E}(x, y, z, t)$$
$$= \mathbf{i}_x E_x(x, y, z, t) + \mathbf{i}_y E_y(x, y, z, t) + \mathbf{i}_z E_z(x, y, z, t)$$

The three components E_x, E_y, E_z are real and vary arbitrarily with time. The field is generated by a collection of sources located at various points in space and having arbitrary time variation. The behaviour of the field is also dependent on the medium through which the fields generated by the sources travelled before reaching the point P. In general, the generation process as well as the properties of the medium are random so all of the field components must be modelled as multidimensional random processes and **E** itself as a vector random process. The statistical characterization of **E** is called *coherence theory*.

Often it is more convenient to treat the field as a complex vector process the real part of which is the actual observed field. Such a situation arises, for example, when any component of the field is of the form

$$|a(t)| \cos[\omega_c t + \phi(t)]$$

and we write the field in the complex form (the *complex field*) as

$$\boldsymbol{a}(t)\, e^{j\omega_c t}, \; \boldsymbol{a}(t) \equiv |a(t)|\, e^{j\phi(t)}$$

A source process responsible for the generation of the field can be modelled as the sum total of infinitely many identical independent elementary processes. Then invoking central limit theorems one can conclude that the source process is gaussian. Gaussian processes

1. The beginner is advised to read §§ 5.2–5 before proceeding with this chapter.

remain gaussian when they propagate through linear time-invariant media. A gaussian process is completely specified if the mean and the correlation function are known. Hence coherence theory deals basically with correlation functions. In summary, the electric field at any point is a complex gaussian process depending on the space coordinates and time as

$$\boldsymbol{E}(P,t) = i_x \boldsymbol{E}_x(x,y,z,t) + i_y \boldsymbol{E}_y(x,y,z,t) + i_z \boldsymbol{E}_z(x,y,z,t) \qquad 3.2$$

3.2 Coherence functions

The autocorrelation function Γ_{xx} of the complex process $\mathbf{E_x}$ is defined as

$$\Gamma_{xx}(P_1, P_2, t_1, t_2) = \mathrm{E}\langle \boldsymbol{E}_x(P_1, t_1)\ \boldsymbol{E}_x^*(P_2, t_2)\rangle \qquad 3.3$$

wherein and hereinafter bold characters shall indicate complex fields and E $\langle \cdots \rangle$ stands for statistical expectation[2]. The autocorrelation functions Γ_{yy} and Γ_{zz} are similarly defined. The cross-correlation Γ_{xy} of two complex processes is defined as

$$\Gamma_{xy}(P_1, P_2, t_1, t_2) = \mathrm{E}\langle \boldsymbol{E}_x(P_1, t_1)\ \boldsymbol{E}_y^*(P_2, t_2)\rangle \qquad 3.4$$

The cross-variance function $\mathbf{C_{xy}}$ of the complex processes $\mathbf{E_x}$ and $\mathbf{E_y}$ is defined by

$$\mathbf{C_{xy}}(P_1, P_2, t_1, t_2)$$

$$\times \mathrm{E}\ \langle [\boldsymbol{E}_x(P_1, t_1) - \mathrm{E}\ \langle\ \boldsymbol{E}_x(P_1, t_1)\rangle] \times [\boldsymbol{E}_y(P_2, t_2) - \mathrm{E}\langle \boldsymbol{E}_y(P_2, t_2)\rangle]\ \rangle$$

or, using eq 3.4, by

$$\mathbf{C_{xy}}(P_1, P_2, t_1, t_2) = \Gamma_{xy}(P_1, P_2, t_1, t_2) - [E\langle \boldsymbol{E}_x(P_1, t_1)\rangle]\ [E\langle \boldsymbol{E}_y(P_2, t_2)\rangle]^* \qquad 3.5$$

The auto-covariance function of the process $\mathbf{E_x}$ is similarly defined. When the mean values of the processes are assumed to be zero the covariance and correlation functions become equal and, without loss of generality, one may assume that the processes are zero-mean as we do in the following. In optics Γ_{xx} is called the coherence function and Γ_{xy} the cross-coherence function.

3.2.1 Properties of the Coherence Function

The following properties of the coherence function may be noted:

$$(i) \quad \Gamma_{xx}(P, P, t, t) = \mathrm{E}\langle \boldsymbol{E}_x(P,t)\, \boldsymbol{E}_x^*(P,t)\rangle > 0 \qquad 3.6$$

$$(ii) \quad |\Gamma_{xx}(P_1, P_2, t_1, t_2)|^2 \leq \Gamma_{xx}\ (P_1, t_1)\ \Gamma_{xx}\ (P_2, t_2) \qquad 3.7$$

The proof of the second property follows from the Schwarz inequality.

2. Do not confuse the E of expectation with the E of the field.

3.2.2 Spatial and Temporal Coherence

At any given time $t \equiv t_1 = t_2$, $\mathbf{E_x}(P_1, t)$ and $\mathbf{E_x}(P_2, t)$ are the complex fields at two spatial points P_1 and P_2, and Γ_{xx} represents the correlation of these two spatially separated fields. Such correlation brings out the spatial variation of the field and is known as *spatial coherence*.

If at a given point the complex fields are measured at two instants of time t_1 and t_2 their correlation $\Gamma_{xx}(P, t_1 ; P, t_2)$ or, more simply written, $\Gamma_{xx}(P, t_1, t_2)$ reflects the relation of the field amplitude at the point P at one instant of time t_1 to the field amplitude at another time instant t_2. In other words, it reflects the field variation with respect to time; this correlation is called *temporal coherence*. If the two types of coherence are separable as

$$\Gamma_{xx}(P_1, t_2; P_2, t_2) = \Gamma_{xx}(P_1, P_2, t) \times \rho\,(t_1, t_2) \qquad 3.8$$

the process $\mathbf{E_x}(P, t)$ is called a spectrally pure process.

3.2.3 Stationary process

We note here some terms and definitions:

A process is called *wide sense stationary* if the coherence function depends only on $\tau \equiv t_1 - t_2$; in that case we have

$$\Gamma_{xx}(P_1, P_2, t_1, t_2) = \Gamma_{xx}(P_1, P_2, \tau) \qquad 3.9$$

If, further, it is spectrally pure we have

$$\Gamma_{xx}(P_1, P_2, \tau) = \Gamma_{xx}(P_1, P_2)\,\rho(\tau) \qquad 3.10$$

A process is called monochromatic if

$$\rho(\tau) = e^{j\omega_0 \tau} \qquad 3.11$$

The power spectral density $S(\omega)$ of the monochromatic process is given by

$$S(\omega) = 2\pi\, u_0(\omega - \omega_0) \qquad 3.12$$

If the time variation of the field is separable and is of the form $\exp(j\,\omega_0\, t)$ the field is called a *harmonic* field. It is clear that a harmonic field is monochromatic but a monochromatic field is not necessarily harmonic.

If ρ is of the form

$$\rho(\tau) = a(\tau)\, e^{j\omega_0 \tau} \qquad 3.13$$

with a slow-varying envelope $a(\tau)$ the field is called *quasi-monochromatic* or *narrow band field*. The power spectrum of such a field is negligible outside a narrow spectral interval around ω_0.

3.2.4 Spatially stationary or Homogeneous Fields

The process is called *spatially stationary* if the coherence function Γ_{xx} depends only on $P_1 - P_2$. Thus

$$\Gamma_{xx}(P+Q,\ Q,\ t+\tau,t) = \Gamma(P)_{xx}\ \rho(\tau) \qquad 3.14$$

3.3 Complex degree of Coherence

The normalized value of the coherence function $\Gamma_{xx}(P_1, P_2, \tau)$ is called the *complex degree of coherence* and denoted by lower case γ:

$$\gamma_{xx}(P_1, P_2, \tau) = \frac{\Gamma_{xx}(P_1, P_2, \tau)}{[\Gamma_{xx}(P_1, P_1, 0)\ \ \Gamma_{xx}(P_2, P_2, 0)]^{\frac{1}{2}}} \qquad 3.15$$

For spectrally pure processes we would have

$$\gamma_{xx}(P_1, P_2, \tau) = \gamma_{xx}(P_1, P_2)\ \rho(\tau) \qquad 3.16$$

3.3.1 Complete incoherence

A process is said to be *completely incoherent* in a region V if

$$\gamma(P_1, P_2) = 0 \ \textit{for} \ P_1 \neq P_2 \qquad 3.17$$

both P_1 and P_2 being in V. In this case we have

$$\Gamma_{xx}(P_1, P_2) = q(P_1)\ \boldsymbol{u}_0(P_1 - P_2) \qquad 3.18$$

where

$$\boldsymbol{u}_0(P_1 - P_2) = \boldsymbol{u}_0(x_1 - x_2)\ \boldsymbol{u}_0(y_1 - y_2)\ \boldsymbol{u}_0(z_1 - z_2) \qquad 3.19$$

The function $q(P)$ is called the *average irradiance*.

3.3.2 Complete coherence

A process is said to be *completely coherent* in a region V if

$$|\gamma_{xx}(P_1, P_2, 0)| \equiv |\gamma_{xx}(P_1, P_2)| = 1 \qquad 3.20$$

for every P_1, P_2 in region V. If a process is coherent then a complex deterministic function $p(P)$ and a random complex signal $\mathbf{Z}(t)$ exist such that

$$E_x(P, t) = p(P)\, Z(t)$$

in the sense that

$$\mathrm{E}\, \langle\, |\, \mathbf{E}_x(P, t) - p(P)\, Z(t)\, |^2\, \rangle = 0$$

Then, for a stationary process, we get

$$\Gamma_{xx}(P_1, P_2, t_1, t_2) = p(P_1)\, p^*(P_2)\, E\, \langle\, Z(t_1)\, Z^*(t_2)\, \rangle$$

$$= p(P_1)\, p^*(P_2) \rho(\tau) \qquad 3.21$$

and also

$$\mathrm{E}\, \langle\, |E_x(P_1, t) + E_x(P_2, t)|^2\, \rangle = |\, p(P_1) + p(P_2)\, |^2 \qquad 3.22$$

The process thus behaves like a deterministic signal with amplitude $p(P)$.

3.3.3 Partial coherence

If the field, by the definitions above, is neither incoherent nor coherent it is said to be *partially coherent.*

3.3.4 Spatially stationary processes

Eq. 3.14 gives the autocorrelation function of a spatially stationary process. The average irradiance of such a process is $\Gamma_{xx}(0)$. If this field were incoherent in the sense defined above we would have

$$\Gamma_{xx}(P) = q\, u_0(P_1 - P_2) \qquad 3.23$$

Note that q is independent of P. If, on the other hand, the field is coherent $p(P)$ is constant:

$$\Gamma_{xx}(P) = \Gamma_{xx}(0) = constant \qquad 3.24$$

3.4 Propagation of the coherence function

Every (scalar) component of the electric field, denoted by **V**, satisfies the wave equation

$$\nabla^2 V - \frac{1}{c^2} \frac{\partial^2 V}{\partial t^2} = -S \qquad 3.25$$

where the source density $\mathbf{S}(P,t)$ is a random process with coherence function given by

$$\Gamma_{ss}(P_1, P_2, t_1, t_2) = E\, \langle\, S(P_1, t_1)\, S^*(P_2, t_2)\, \rangle \qquad 3.26$$

Taking the conjugate of eq. 3.24, evaluating at P_2, t_2, multiplying it by $\mathbf{S}(P_1,t_1)$ and taking the expectation, the following equation governing the cross-coherence function is obtained:

$$\nabla_2^{\,2}\,\Gamma_{sv} - \frac{1}{c^2}\frac{\partial^2 \Gamma_{sv}}{\partial t_2^2} = -\,\Gamma_{ss} \qquad 3.27$$

Thus the cross-coherence function between source $\mathbf{S}(P_1,t_1)$ at (P_1,t_1) and the resulting field $\mathbf{V}(P_2,t_2)$ at variable point (P_2,t_2) satisfies the wave equation with a deterministic source function Γ_{ss}. Similarly we have the equation

$$\nabla_1^{\,2}\,\Gamma_{vv} - \frac{1}{c^2}\frac{\partial^2 \Gamma_{vv}}{\partial t_1^2} = -\,\Gamma_{sv} \qquad 3.28$$

If the source coherence function Γ_{ss} is stationary with respect to time then the same is true of Γ_{sv} and Γ_{vv}, resulting in

$$\nabla_2^{\,2}\,\Gamma_{sv} - \frac{1}{c^2}\frac{\partial^2 \Gamma_{sv}}{\partial \tau^2} = -\,\Gamma_{ss} \qquad 3.29$$

$$\nabla_1^{\,2}\,\Gamma_{vv} - \frac{1}{c^2}\frac{\partial^2 \Gamma_{vv}}{\partial \tau^2} = -\,\Gamma_{sv} \qquad 3.30$$

3.5 Propagation of coherence functions throughs Linear Systems

Figure 3.1 shows the system model[3], $\mathbf{X}(P_1, t_1')$ and $\mathbf{Y}(Q_1,t_1)$ being the input and output respectively. If $\mathbf{X}$ and $\mathbf{Y}$ are connected as

$$\boldsymbol{Y}(Q_2, t_2) = \boldsymbol{X}(P_2, t_1') \otimes \boldsymbol{h}(P_2, Q_2, t_2, t_1') \qquad 3.31$$

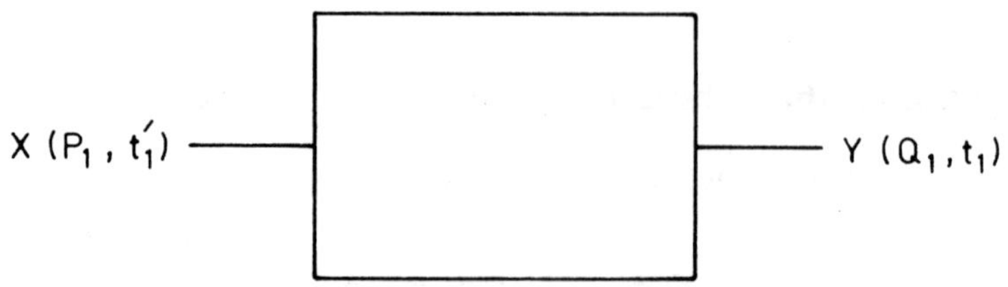

Figure 3.1 Linear system model for coherence function propagation

3. See Chapter 4 for other kinds of system model.

we have

$$\Gamma_{xy}(P_1, Q_2, t_1, t_2) = \mathrm{E}\,\langle \boldsymbol{X}(P_1, t_1)\ \boldsymbol{Y}^*(Q_2, t_2) \rangle$$

$$= \Gamma_{xx}(P_1, P_2, t_1, t_2) \otimes \boldsymbol{h}^*(P_2, Q_2, t_2, t'_1) \qquad 3.32$$

In the same way we get

$$\Gamma_{yy}(Q_1, Q_2, t_1, t_2) = \mathrm{E}\,\langle \boldsymbol{Y}(Q_1, t_1)\ \boldsymbol{Y}^*(Q_2, t_2) \rangle$$

$$= \mathrm{E}\,\langle \boldsymbol{Y}(Q_1, t_1)\,[\boldsymbol{X}^*(P_2, t'_1) \otimes \boldsymbol{h}^*(P_2, Q_2, t_2, t'_1)] \rangle$$

which, using eq. 3.31-32, may be written as

$$\Gamma_{yy}(Q_1, Q_2, t_1, t_2) = \boldsymbol{h}(P_1, Q_1, t'_2, t_1) \otimes [\Gamma_{xx}(P_1, P_2, t'_2, t'_1)$$

$$\otimes\ \boldsymbol{h}^*(P_2, Q_2, t_2, t'_1)] \qquad 3.33$$

These results can be conveniently used to obtain the coherence functions of diffracted fields.

3.6 Generalization

The generalized coherence function of the vector field at two fixed points (P_1,t_1) and (P_2,t_2) is defined as

$$\mathrm{E}\,\langle \boldsymbol{E}(P_1, t_1)\ \boldsymbol{E}^\dagger(P_2, t_2) \rangle = \begin{bmatrix} \Gamma_{xx} & \Gamma_{xy} & \Gamma_{xz} \\ \Gamma_{yx} & \Gamma_{yy} & \Gamma_{yz} \\ \Gamma_{zx} & \Gamma_{zy} & \Gamma_{zz} \end{bmatrix} \qquad 3.34$$

where † stands for the complex conjugate transpose and $\mathbf{E}(P,t)$ is the column vector

$$\boldsymbol{E}(P,t) = \begin{bmatrix} \boldsymbol{E}_x(P, t) \\ \boldsymbol{E}_y(P, t) \\ \boldsymbol{E}_z(P, t) \end{bmatrix}$$

Further generalizations are possible for defining the coherence and cross-coherence functions of all of the components of electric and magnetic fields.

3.7 Polarization

If the field configuration pertains to a plane wave propagating in the z-direction the electric (and magnetic) field(s) is (are) restricted to remain in the transversal xy-plane and has (have), in general, both $\mathbf{E}_x$ and $\mathbf{E}_y$ ($\mathbf{B}_x$ and $\mathbf{B}_y$) components. For a plane wave the field components are independent of x and y but are functions of time. The resulting electric (magnetic) field vector traces a trajectory in the *xy*-plane as time *t* varies. For harmonic signals, depending on the relative amplitudes and phases of $\mathbf{E}_x$ and $\mathbf{E}_y$, the trajectory is, in general, an ellipse of either handedness. It degenerates into a coincident pair of straight lines of positive or negative slope, or a circle of either handedness for specific ratios/values of them. Accordingly the field is said to be elliptically, linearly or circularly polarized.[4] However, when the two components $\mathbf{E}_x$ and $\mathbf{E}_y$ are not completely coherent the trajectory varies randomly.

Paralleling the discussion of partial coherence (§3.3.3), *partial polarization* may be defined to describe such situations. It is important to note that the concept of partial polarization is intimately connected with that of partial coherence.

3.7.1 Coherency matrix

The coherency matrix J is defined as

$$J = \mathrm{E}\,\langle \boldsymbol{E}\boldsymbol{E}^{\dagger} \rangle \tag{3.35}$$

where $$\boldsymbol{E} = \begin{bmatrix} E_x \\ E_y \end{bmatrix}$$

The matrix J can be diagonalized; the diagonal elements are the non-negative values λ_1 and $\lambda_2 < \lambda_1$. The diagonalized matrix may be broken down as the sum of two diagonal matrices:

$$\begin{bmatrix} \lambda_1 & 0 \\ 0 & \lambda_2 \end{bmatrix} = \begin{bmatrix} \lambda_2 & 0 \\ 0 & \lambda_2 \end{bmatrix} + \begin{bmatrix} \lambda_1 - \lambda_2 & 0 \\ 0 & 0 \end{bmatrix} \tag{3.36}$$

The first represents the unpolarized component and the second the polarized one. Such a decomposition is unique. The degree of polarization P is defined as the ratio of the irradiance of the polarized component to the total irradiance and is given by

$$P = \frac{\lambda_1 - \lambda_2}{\lambda_1 + \lambda_2}$$

$$= \left[1 - \frac{4\,(\det J)}{(Tr\,J)^2}\right]^{\frac{1}{2}} \tag{3.37}$$

P lies between 0 and l; the former value corresponds to completely incoherent radiation and the latter to completely coherent radiation. $0 < P < 1$ corresponds to partially coherent radiation.

---

4. See § 5.2 for a less formal description of polarization.

3.7.2 Stokes parameters

The Stokes parameters of polarized radiation defined originally in terms of the real transverse components E_x, E_y and the phase difference δ between them as

$$\left.\begin{aligned} S_0 &= E_x^2 + E_y^2 \\ S_1 &= E_x^2 - E_y^2 \\ S_2 &= 2\,E_x\,E_y \cos\delta \\ S_3 &= 2\,E_x\,E_y \sin\delta \end{aligned}\right] \tag{3.38}$$

are related to the components of J as follows:

$$\begin{bmatrix} S_0 \\ S_1 \\ S_2 \\ S_3 \end{bmatrix} = \begin{bmatrix} 1 & 0 & 0 & 1 \\ 1 & 0 & 0 & -1 \\ 0 & 1 & 1 & 0 \\ 0 & -j & j & 0 \end{bmatrix} \begin{bmatrix} J_{xx} \\ J_{xy} \\ J_{yx} \\ J_{yy} \end{bmatrix} \tag{3.39}$$

In terms of the Stokes parameters the degree of polarization is given by

$$P = \frac{\sqrt{(S_1^2 + S_2^2 + S_3^2)}}{S_0} \tag{3.40}$$

Another representation of polarized light is the Poincaré sphere for which (and for ways of relating Stokes parameters to which) reference may be made to the sources cited at the end of this chapter.

3.7.3 Transmission of coherency matrix through optical elements

Let an optical element have a scattering matrix [T] such that the input (sub-subscript *i*) and output (sub-subscript *o*) fields are related by

$$\begin{bmatrix} \boldsymbol{E}_{x_0} \\ \boldsymbol{E}_{y_0} \end{bmatrix} = [T] \begin{bmatrix} \boldsymbol{E}_{x_i} \\ \boldsymbol{E}_{y_i} \end{bmatrix} \tag{3.41}$$

whereby we have

$$\begin{aligned} J_0 &= \mathrm{E}\langle \boldsymbol{E}_0 \boldsymbol{E}_0^{\dagger} \rangle \\ &= \mathrm{E}\langle T\boldsymbol{E}_i \boldsymbol{E}_i^{\dagger} \boldsymbol{T}^{\dagger} \rangle = T J_i\, T^{\dagger} \end{aligned} \tag{3.42}$$

If the optical element is lossless we have

$$Tr\,[J_0] = Tr\,[J_i\,] \tag{3.43}$$

and *vicé versâ*.

3.7.4 Muller Matrices

The transmission of Stokes parameters through optical elements can equivalently be described by *Mueller matrices* [M] defined as

$$[S_0] \equiv [M][S_i] \qquad 3.44$$

Mueller matrices for commonly occurring optical elements were tabulated by Mueller and may be seen in the references cited at the end of this chapter. The relation between the scattering matrix and the Mueller matrix may easily be derived (*loc cit*).

3.8 Summary

In this chapter we have given a concise account of how a formalism based on statistical concepts can be developed to deal with what generally vaguely is called "coherence" and with its propagation through linear systems, using the system model concept of electrical circuit theory. At the same time it has been our attempt to bring out the generally neglected close connection between not only coherence and polarization but also coherency and polarization matrices.

Bibliography

BORN Max and Wolf Emil, 1975 *Principles of Optics* 5th ed (Oxford: Pergamon Press) Chapters VII and X.

GOODMAN Joseph W, 1985 *Statistical Optics* (New York: Wiley-Interscience).

MARĀTHAY Arvind S, 1982 *Elements of Optical Coherence Theory* (New York: John Willey and Sons).

SHURCLIFF William A, 1966 *Polarized Light* (Cambridge, Mass: Harvard University Press) Chapters 1, 2, 8 and Appendix 2.

SHURCLIFF William A and Ballard Stanley S, 1964 *Polarized Light*, a Van Nostrand Momentum Book #7 (Princeton: D Van Nostrand Co) for Jones and Mueller matrices.

Chapter 4

OPTICAL SIGNAL PROCESSING

K R SARMĀ

Introduction

Optical elements have been used for centuries to process optical signals. However, it is only recently that optical signal processing (*OSP*) techniques are considered from the viewpoint of system theory. The impetus to this came from the discovery of the Fourier-transforming property of a lens and the availability of coherent sources of light, *viz* lasers. Before discussing the techniques of optical signal processing a review of linear system theory is in order not only for the purpose of introducing physicists to somewhat unfamiliar concepts but also because optical signal processing is based on an extension of the one-dimensional theory to two dimensions. The difference is that the "one dimension" of communication theory is *time* whereas the two dimensions of OSP are *spatial*.

4.1 Linear system theory

4.1.1 Impulse response

Let us consider a one-dimensional system with input $x(t)$ and output $y(t)$ related by $y(t) = L[x(t)]$. The system is called *linear* if

$$L[\alpha x_1 + \beta x_2] = \alpha L[x_1] + \beta L[x_2] \qquad 4.1$$

for any arbitrary inputs x_1 and x_2 and constants α and β The system is called *shift-invariant* if

$$L[x(t + \tau)] = y(t + \tau) \qquad 4.2$$

Linear shift-invariant systems are completely specified by the *impulse response h(t)* which is the response of the system to a unit impulse input.

A unit impulse $h_0(t)$ is defined by the two properties

$$\int_{-\infty}^{\infty} h_0(t)\, dt = 1 \qquad 4.3$$

$$\int_{-\infty}^{\infty} f(\tau)\ h_0(t-\tau)\ d\tau = f(t) \qquad 4.4$$

Given the impulse response $h(t)$, the response $y(t)$ to an arbitrary input $x(t)$ can be obtained as follows:

$$u_0(t) \rightarrow h(t)$$

$$u_0(t-\tau) \rightarrow h(t-\tau)\ [shift\text{– }invariance]$$

$$x(\tau)\ u_0(t-\tau) \rightarrow x(\tau)\ h(t-\tau)\ [linearity]$$

$$x(t) = \int_{-\infty}^{\infty} x(\tau)\ u_0(t-\tau) = \int_{-\infty}^{\infty} h(\tau)\ x(t-\tau)\, d\tau$$

so that

$$y(t) = \int_{-\infty}^{\infty} x(\tau)\ h(t-\tau)\, d\tau$$

$$= \int_{-\infty}^{\infty} h(\tau)\ x(t-\tau)\, d\tau \qquad 4.5$$

This integral is called the *convolution integral* (*see* Chapter 1). Hence for a linear shift-invariant system the input and output are related through the convolution integral. As in Chapter 1 we denote it as

$$y = h \otimes x \qquad 4.6$$

4.1.2 Response to exp (jω t)

Substituting $x(t) = \exp(j\,\omega\,t)$ into eq 4.6 we get

$$y(t) = \int_{-\infty}^{\infty} h(\tau)\ e^{j\omega(t-\tau)}\ d\tau$$

$$= \left[\int_{-\infty}^{\infty} h(\tau)\ e^{-j\omega\tau}\ d\tau\right] e^{j\omega t} \qquad 4.7$$

$$\equiv H(\omega)\ e^{j\omega t}$$

where the integral within the square brackets is denoted by the symbol H(ω). Clearly H(ω) is the Fourier transform of the impulse response. Thus the response of a linear shift-invariant system to exp $(j\,\omega\,t)$ is exp $(j\,\omega\,t)$ weighted by H(ω), the Fourier transform of the impulse response. H(ω) is called the transfer function. It is worthwhile noting that

$$L\,[e^{j\omega t}] = \lambda\, e^{j\omega t}, \text{ where } \lambda = H(\omega) \qquad 4.8$$

Hence exp $(j\,\omega\,t)$ is an eigenfunction of a linear shift-invariant system. As a result the analysis of the system becomes easier if the input is expressed as a weighted sum of the eigenfunctions exp $(j\,\omega\,t)$, *i. e.* if the input is represented by its Fourier representation

$$x(t) = \frac{1}{2\pi}\int_{-\infty}^{\infty} X(\omega)\, e^{j\omega t}\, d\omega \qquad 4.9$$

4.1.3 Relationship in frequency domain

Taking Fourier transform of the convolution relationship eq 4.6 and using the convolution theorem, eq 1.16, we get

$$Y(\omega) = H(\omega)\cdot X(\omega) \qquad 4.10$$

Thus the FT of the output is equal to the product of the FT of the input and the transfer function. The output *Y(t)* can then be obtained by taking the inverse FT of *Y* (ω):

$$y(t) = \frac{1}{2\pi}\int_{-\infty}^{\infty} Y(\omega)\, e^{j\omega t}\, d\omega \qquad 4.11$$

4.1.4 Two-dimensional systems

Optical systems process signals which are functions of spatial variables *x* and *y*. An input *f*(*x*,*y*) gives rise to an output *g*(*x*,*y*). Optical elements can be modelled as linear shift-invariant systems as we shall see later. A two-dimensional shift-invariant system is completely characterized by an impulse response *h*(*x*,*y*) which is the output of the system due to an impulse located at $x, y = 0$. Following the arguments above for one-dimensional systems the input and the output are related by

$$g(x, y) = \int_{-\infty}^{\infty}\int_{-\infty}^{\infty} f(x_0, y_0)\; h(x - x_0\,,\; y - y_0)\; dx_0\, dy_0 \qquad 4.12$$

Taking FT on both sides we get

$$G(\omega_x\,, \omega_y) = H(\omega_x\,, \omega_y)\cdot F(\omega_x\,, \omega_y) \qquad 4.13$$

where G is the Fourier transform of g:

$$G(\omega_x , \omega_y) = \int_{-\infty}^{\infty}\int_{-\infty}^{\infty} g(x, y)\; e^{-j(\omega_x x + \omega_y y)}\, dx\, dy \qquad 4.14$$

As the variables x,y are spatial the *spatial frequencies* ω have the dimensions of inverse length. A convenient unit for spatial frequency is *lines mm*$^{-1}$.

4.1.5 Techniques of analysis

Linear shift-invariant systems can be analyzed either in the spatial domain *(x,y)* through the convolution integral or equivalently in the frequency domain by the product relation. The former method is called *space domain analysis* and the latter *frequency domain analysis*. Although it may appear that the frequency domain analysis is a roundabout way it is easier to follow when the signals are piecewise continuous.

4.2 Models of optical elements

The three basic elements in any optical system are free space, lens and film transparency. We obtain the system models for these elements in §§4.2.1-3.

4.2.1 Free space

It has been shown that under Fresnel approximation the field $E(x,y,z)$ at any point on a plane at $z=z$ is related to the field $E(x_0,y_0,0)$ on the plane at $z=0$ through the integral relation given by

$$E(x, y, z) = \frac{j\, e^{-jkz}}{\lambda z} \int_{-\infty}^{\infty}\int_{-\infty}^{\infty} E(x_0, y_0, 0)\; e^{\frac{-jk}{2z}[(x - x_0)^2 + (y - y_0)^2]}\, dx_0\, dy_0 \qquad 4.15$$

An alternative derivation is as follows. The field at any point P(x,y,z) may be viewed as a sum of the contributions from an infinite set of plane waves travelling in arbitrary directions:

$$E(x, y, z) = \int_{-\infty}^{\infty}\int_{-\infty}^{\infty} F(\alpha, \beta)\; e^{-jk(\alpha x + \beta y + \gamma z)}\, d\alpha\, d\beta \qquad 4.16$$

where α , β , γ are the direction cosines of the propagation vector of a wave, so that $\alpha^2 + \beta^2 + \gamma^2 = 1$, and j F (α , β) is the amplitude of that wave. F is called the *plane wave spectrum* or *angular spectrum* or *wave number spectrum* of $E(x,y,z)$. Evaluating the field at $z = 0$ we get

$$E(x, y, 0) = \int_{-\infty}^{\infty}\int_{-\infty}^{\infty} F(\alpha, \beta)\, e^{-jk(\alpha x + \beta y)}\, d\alpha\, d\beta$$

Noting that this is Fourier relationship we invert it to get

$$F(\alpha, \beta) = \frac{1}{\lambda^2}\int_{-\infty}^{\infty}\int_{-\infty}^{\infty} E(x_0, y_0, 0)\, e^{jk(\alpha x_0 + \beta y_0)}\, dx_0\, dy_0$$

so that

$$E(x, y, z) = \frac{1}{\lambda^2}\int_{-\infty}^{\infty}\int_{-\infty}^{\infty}\int_{-\infty}^{\infty}\int_{-\infty}^{\infty} E(x_0, y_0, 0)\, e^{-jk[\alpha(x - x_0) + \beta(y - y_0)]}\, d\alpha\, d\beta\, dx_0\, dy_0$$

Putting $\gamma = \sqrt{[1 - (\alpha^2 + \beta^2)]}$ and making the usual assumption in Fresnel diffraction that $|\alpha|, |\beta| << 1$, γ may be approximated as $[1 - (\alpha^2 + \beta^2)/2]$. Substituting in the above and integrating over α and β we get the Fresnel integral. Alternatively we may rewrite eq 4.16 with Fresnel approximation as

$$E(x, y, z) = e^{-jkz}\int_{-\infty}^{\infty}\int_{-\infty}^{\infty} F(\alpha, \beta)\, e^{\frac{jk}{2}(\alpha^2 + \beta^2)}\, e^{-jk(\alpha x + \beta y)}\, d\alpha\, d\beta$$

Noting that the integral is the FT of the product of the two functions F(α, β) and the first exponential in the integrand and remembering that the FT of the product of two functions is the convolution of their individual FTs we obtain

$$E(x, y, z) = \frac{j e^{-jkz}}{\lambda z}\int_{-\infty}^{\infty}\int_{-\infty}^{\infty} E(x_0, y_0, 0)\, e^{\frac{-jk}{2z}[(x - x_0)^2 + (y - y_0)^2]}\, dx_0\, dy_0$$

which is the same as eq 4.15. This Fresnel integral may be viewed as the response *E(x,y,z)* of a linear shift-invariant system with impulse response *h(x,y)* and input $E(x_0, y_0, 0)$ where we put

$$h(x, y) = \frac{j e^{-jkz}}{\lambda z}\, e^{-\frac{jk}{2z}(x^2 + y^2)} \qquad 4.17$$

Note that the impulse response *h(x,y)* denotes a spherical wavefront produced by a point source located at (0,0,0). Thus the field *E(x,y,z)* at the point P is the sum of such spherical waves generated by treating every point on the plane $z = 0$ as a point source located at that

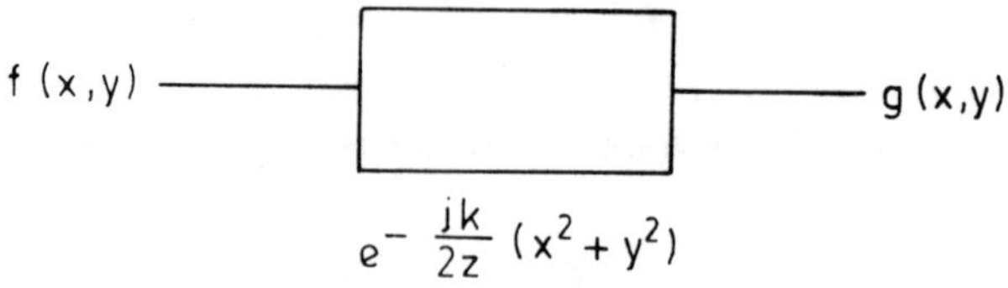

Figure 4.1 System model for free space

point. Normally we ignore the constant exponential factor $[\exp(-jkz)]/\lambda z$ in h and normalize h as

$$h(x, y) = e^{-\frac{jk}{2z}(x^2 + y^2)} \tag{4.18}$$

for "a piece of semi-infinite free space of length z". The system model is shown in Figure 4.1.

4.2.2 Lens

The lens is the other most commonly used optical element. It introduces a differential phase shift to the field at various points due to its geometric shape and the refractive index (n) of its material, here assumed uniform throughout its body. Let us calculate the transmission properties of the lens. Figure 4.2 shows a lens with a thickness Δ_0 at the centre and tapering off towards the circular edge depending upon the geometrical shape. In the figure are indicated the input plane with a field distribution $f(x,y)$ and the output plane with a field distribution $g(x,y)$. The phase introduced by the propagation through a distance comprising free space δ_1, lens thickness $\Delta(x, y)$ and free space δ_2 is given by

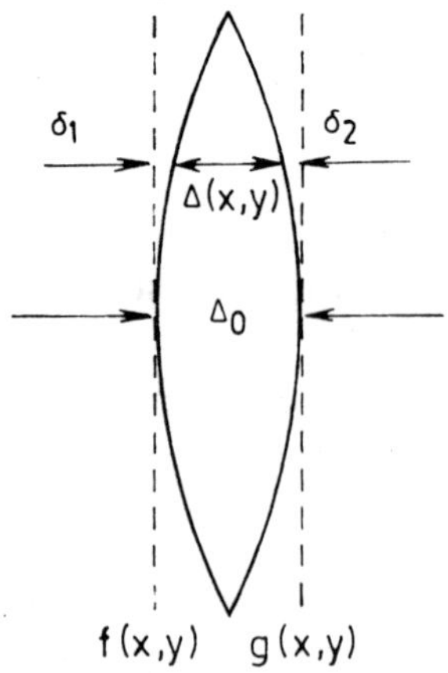

Figure 4.2 Input and output fields of a convex lens with lens function $\Delta(x,y)$ and central thickness Δ_0

$$\phi(x, y) = k(\delta_1 + \delta_2) + n k \Delta(x, y)$$

or, considering that $\delta_1 + \delta_2 + \Delta(x, y) = \Delta_0$,

$$\phi(x, y) = n k \Delta_0 - (n-1) k (\delta_1 + \delta_2)$$

Thus the output field distribution is

$$g(x, y) = f(x, y)\, e^{-j\phi(x, y)}$$

$$= f(x, y)\, e^{-jnk\Delta_0}\, e^{j(n-1)k(\delta_1 + \delta_2)}$$

The constant factor $\exp(-jnk\Delta_0)$ may be ignored. In optics a standard sign convention is adopted which is as follows. When a ray travels from the left to the right a convex surface encountered has a positive index and a concave surface a negative one. For a thin lens it can be deduced that

$$\delta_1 + \delta_2 = \frac{(x^2 + y^2)}{2}\left[\frac{1}{R_1} - \frac{1}{R_2}\right]$$

where R_1 and R_2 are the radii of curvature of the input and output surfaces of the lens respectively with their polarities decided by the sign convention. The focal length of the lens is defined as

$$\frac{1}{F} = (n-1)\left[\frac{1}{R_1} - \frac{1}{R_2}\right] \qquad 4.19$$

Hence the phase factor associated with the lens of focal length F is given by

$$e^{\frac{jk}{2F}(x^2 + y^2)} \qquad 4.20$$

and the lens acts as a multiplier with this multiplying function. The system model is shown in Figure 4.3.

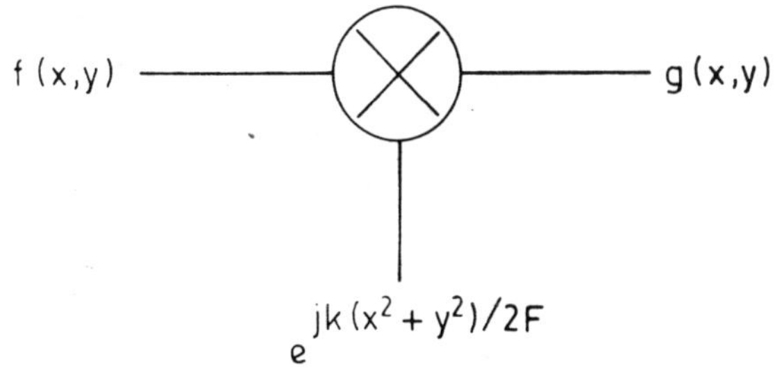

Figure 4.3 System model for a convex lens

4.2.3 Film transparency

Transparencies are used as inputs or filters in optical processing systems. The transmittance $T(x,y)$ of a transparency varies from point to point and is a non-negative real number. The input and output fields of a transparency are related by the relation

$$g(x, y) = T(x, y)\, f(x, y) \tag{4.21}$$

The system model is shown in Figure 4.4.

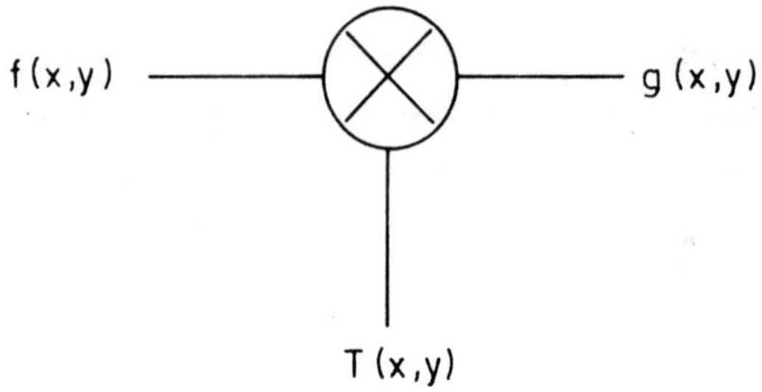

Figure 4.4 System model for a film transparency

4.3 Basic optical system

A basic optical processing system consisting of a single lens is shown in Figure 4.5. The input field $l(u,v)$ on the u,v-plane is given by

$$l(u, v) = f(x, y) \otimes h_1(x, y)$$

$$= \frac{1}{\lambda D_1} \int_{-\infty}^{\infty} \int_{-\infty}^{\infty} f(x, y)\, e^{\frac{-jk}{2D_1}[(u-x)^2 + (v-y)^2]}\, e^{\frac{jk}{2F}(x^2 + y^2)}\, dx\, dy$$

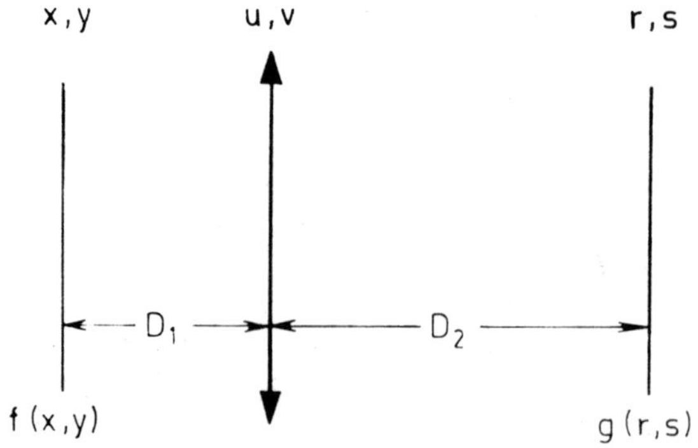

Figure 4.5 Basic optical system consisting of a single lens

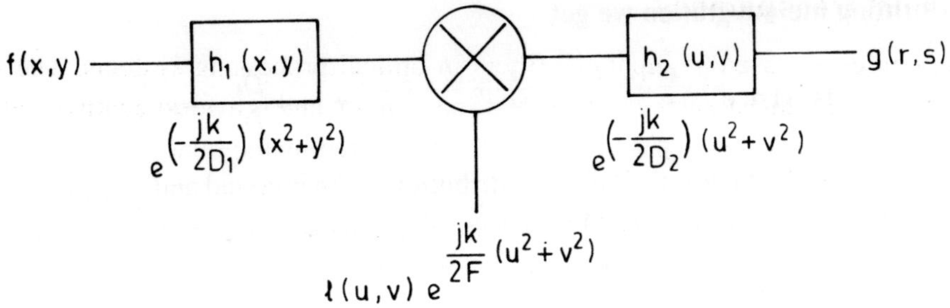

Figure 4.6 System model for Figure 4.5

We assume an infinite aperture; the effects of a finite aperture will be dealt with later. Following the system model shown in Figure 4.6 we obtain the field distribution *g(r,s)* at the output plane as

$$g(r, s) = \frac{j}{\lambda D_2} \int_{-\infty}^{\infty}\int_{-\infty}^{\infty} l(u, v)\, e^{j\frac{k}{2F}(u^2+v^2)}\, e^{-j\frac{k}{2D_2}\left[(u-r)^2+(v-s)^2\right]}\, du\, dv$$

Substituting for *l(u,v)* and simplifying we get

$$g(r, s) = -\frac{e^{-j\frac{k}{2D_2}(r^2+s^2)}}{\lambda^2 D_1 D_2} \int_{-\infty}^{\infty}\int_{-\infty}^{\infty}\int_{-\infty}^{\infty}\int_{-\infty}^{\infty} f(x, y)\, e^{-j\frac{k}{2D_1}(x^2+y^2)}\, e^{\frac{-k}{2}(u^2+v^2)\left[\frac{1}{D_1}+\frac{1}{D_2}-\frac{1}{F}\right]}$$

$$\times\, e^{j\frac{k}{D_1}(ux+vy)}\, e^{j\frac{k}{D_2}(ru+sv)}\, du\, dv\, dx\, dy$$

4.3.1 Imaging condition

If the distances D_1 and D_2 are such that

$$\frac{1}{D_1} + \frac{1}{D_2} = \frac{1}{F} \qquad 4.22$$

the field distribution at the output becomes

$$g(r, s) = -\frac{\exp\left[-j\,(k/2)\; D_2(r^2+s^2)\right]}{\lambda^2 D_1 D_2} \int_{-\infty}^{\infty}\int_{-\infty}^{\infty} f(x, y)\, \exp\left[-j\,(k/2)\; D_1(x^2+y^2)\right]$$

$$\times \left\{ \int_{-\infty}^{\infty}\int_{-\infty}^{\infty} \exp j\,(k/D_1)\left[\, u\,(x+(D_1/D_2)\, r\,) + v\,(\,y+(D_1/D_2\,)\, s)\,\right] du\, dv \right\} dx\, dy$$

On performing the integration we get

$$g(r,s) \propto e^{-j\frac{k}{2D_2}(r^2+s^2)}\; e^{-j\frac{k}{2D_1}\left(r^2\frac{D_1^2}{D_2^2}+s^2\frac{D_1^2}{D_2^2}\right)} f\left(-r\frac{D_1}{D_2},\, -s\frac{D_1}{D_2}\right) \qquad 4.23$$

Thus, but for a phase factor, the output distribution is the inverted and scaled version of the input distribution, magnified or reduced accordingly as $D_1 > D_2$ or $D_1 < D_2$. The geometrical optical derivation—by any method—leads only to eq 4.22; it does not concern itself with the phase problem.

4.3.2 Fourier transforming condition

If $D_1=D_2=F$ we have

$$g(r,s) \propto \int_{-\infty}^{\infty}\int_{-\infty}^{\infty} f(x,y)\, e^{j\frac{k}{F}(xr+ys)}\, dx\, dy \qquad 4.24$$

which is the Fourier transform of the input distribution. A detailed analysis of this will be taken up later. The discovery of the Fourier transforming property of the lens led to the extensive use of optical methods in signal processing.

4.3.3 Operational notation in optical system analysis

A commonly occurring function in analyzing the optical system is the Ψ-function given by

$$\Psi(x,y,d) \equiv e^{-j\frac{kd}{2}(x^2+y^2)} \qquad 4.25$$

The following properties of the Ψ -function may easily be established:

$$\Psi(-x,-y,d) = \Psi(x,y,d)$$

$$\Psi(x,y,-d) = \Psi^*(x,y,d)$$

$$\Psi(x,y,d_1) \times \Psi(x,y,d_2) = \Psi(x,y,d_1+d_2)$$

$$\Psi(x-u,y-v,d) = \Psi(u,v,d)\ \Psi(x,y,d)\ e^{jkd(xu+yv)}$$

$$\Psi(x,y,d) = \Psi(x,0,d) \times \Psi(0,y,d)$$

$$\Psi(x,y,0) = 1$$

$$\lim_{d\to\infty} d\,\Psi(x,y,d) = u_0(x,y) \qquad 4.26$$

and, finally,

$$\int_{-\infty}^{\infty}\int_{-\infty}^{\infty} \Psi(x, y, d_1)\, e^{jkd^2(ux+vy)}\, dx\, dy = \frac{(const)}{d_1}\, \Psi^*\left(u, v, \frac{d_2^{\,2}}{d_1}\right)$$

The use of the operational notation simplifies the manipulation of formulae in the analysis of optical systems as we shall see in the following sections.

4.3.4 Basic optical system analysis using operational notation

As an example, consider the system shown in Figure 4.7a,b. Using the operational notation we write

$$g(r, s) = \{[f(x, y) \otimes h_1(x, y)]\, \Psi^*(u, v, f)\} \otimes h_2(u, v)$$

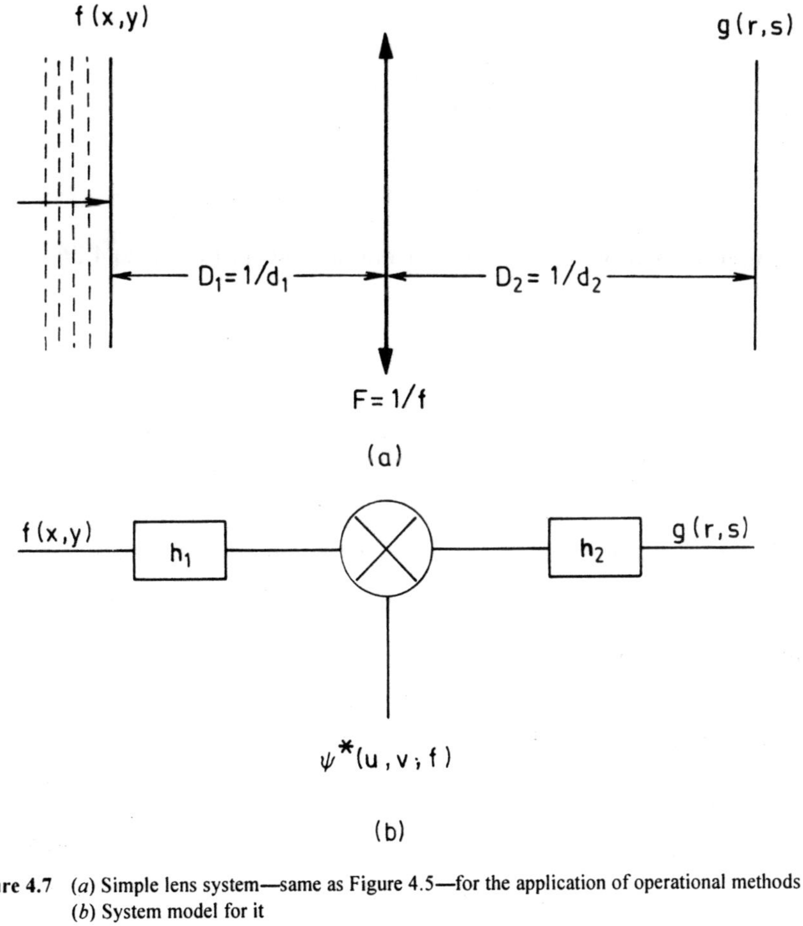

Figure 4.7 (*a*) Simple lens system—same as Figure 4.5—for the application of operational methods (*b*) System model for it

$$= \{[f(x,y) \otimes \Psi(x,y,d_1)] \times \Psi^*(u,v,f)\} \otimes \Psi(u,v,d_2)$$

$$= \int_{-\infty}^{\infty} f(x,y)\ \Psi\ (x-u, y-v, d_1)\ \Psi^*(u,v,f)\ \Psi(r-u,\ s-v, d_2)\ d\,(u,v,x,y)$$

Working out step by step we get

$$g\,(r,s)$$

$$= \Psi(r,s,d_2)$$

$$\times \int_{-\infty}^{\infty}\int_{-\infty}^{\infty} f(x,y)\ \Psi(x,y,d_1) \int_{-\infty}^{\infty}\int_{-\infty}^{\infty} \Psi(u,v,\ d_1+d_2-f)$$

$$\times\ \exp\{\,j\,k\,d_1\,[u\,(x+(d_2\,/\,d_1)\,r)+v\,(\,y+(d_2\,/\,d_1)\,s)]\}\ du\ dv\ dx\ dy$$

Integrating with respect to u and v,

$$\Psi(r,s,d_1) = \int_{-\infty}^{\infty}\int_{-\infty}^{\infty} f(x,y)\ \Psi(x,y,d_1\,)$$

$$\times\Psi^*\left(x\ +\ (d_2\,/\,d_1)\,r,\ y\ +\ (d_2\,/\,d_1)\,s,\ d_1^{\ 2}\,/\,(d_1+d_2-f)\right)\ dx\,dy$$

Using the imaging condition, $d_1+d_2-f = 0$ and the last relation in eq 4.26,

$$g\ \ (r,s)$$

$$= \Psi\,(r,s,d_2) \int_{-\infty}^{\infty}\int_{-\infty}^{\infty} f(x,y)\ \Psi(x,y,d_1)\ u_0\,[\,x+(d_2\,/\,d_1)\,r, y+(d_2\,/\,d_1)\,s\,]\,dx\,dy$$

$$= \Psi\ [\,r,s,d_2+(d_2^{\ 2}\,/\,d_1)\,]\ f[\,-(d_2\,/\,d_1)\,r,\ -(d_2\,/\,d_1)\,s\,] \tag{4.27}$$

Thus we get an inverted and scaled image—a condition arrived at earlier by straightforward calculation.

If we take $d_1 = d_2 = f$ we get

$$g\,(r,s) = \Psi\,(r,s,f) \int_{-\infty}^{\infty}\int_{-\infty}^{\infty} f(x,y)\ \Psi(x,y,f)$$

$$= \Psi^*\ (x,y,f)\ \Psi^*\ (r,s,f)\ e^{j\,k\,f(r\,x\,+\,s\,y)}\ d\,x\,d\,y$$

Using the results of the operational method we have

$$g(r,s) = \int_{-\infty}^{\infty}\int_{-\infty}^{\infty} f(x,y)\, e^{jkf(rx+sy)}\, dx\, dy$$

As seen in eq 4.24, the output is the Fourier transform of the input. Comparing this with eq 2.32 which established that the Fraunhofer diffraction pattern was related to the Fourier transform of the input distribution, we see that under the FT condition, *viz* that both the input and output planes be at distances equal to the focal length of the lens on the opposite sides, we get an *exact* FT; that is, without the constant factor in front of the integral as we got in eq 2.32.

4.4 Spatial filtering

Figure 4.8 shows a basic optical system for yet another purpose. The first lens (of focal length F_1) gives at the u,v-plane the Fourier transform of the input $f(x,y)$ which is FT-ed by the second lens (of focal length F_2). This double transforming should obviously give $f(x,y)$ back—provided that the component $H(u,v)$ is absent. This leads to an important technique as we now show. $H(u,v)$ is the transmittance function of a *complex mask* placed at the FT-plane of the first lens. The mask is both absorptive and retardive. The output of the lens is $FT\{f(x,y)\} \equiv F(u,v)$ so the output of the mask is the product of $F(u,v)$ and $H(u,v)$ which is re-transformed by the second lens. What the mask does is to modify the FT and then send it for further transformation to the second lens. This is called *spatial filtering* in analogy with similar time-frequency-domain operation of filtering in electrical circuits.

In simple filtering schemes like lowpass, highpass, *etc* apertures or stops may be used in the u,v-plane to represent $H(u,v)$. For example, if a stop is placed at the origin of the

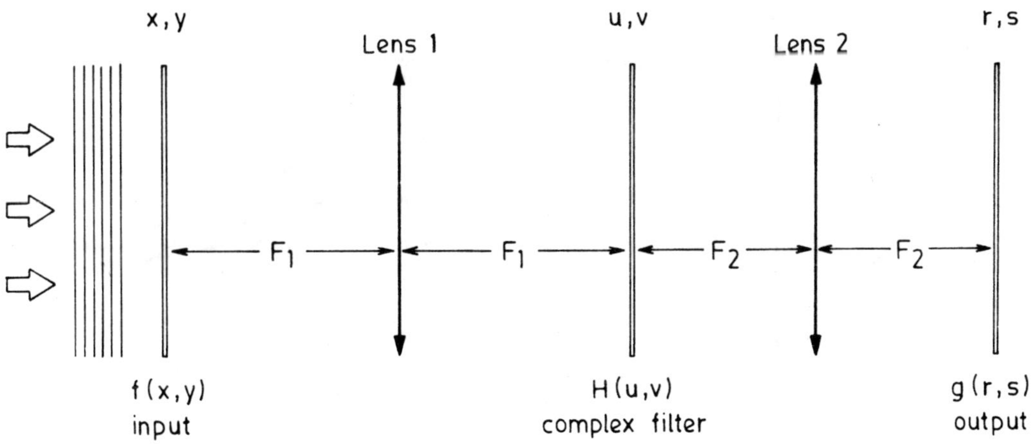

Figure 4.8 Illustrating FT property of a convex lens

u, v-plane we will get an output with its low frequency content filtered out and the high frequency content retained. In general, however, a complex mask is required. For example, in the case of matched filtering for either the detection of a signal in additive gaussian noise or in pattern recognition by correlation the transfer function of the filter is $S^*(u,v)$ where $S(u,v)$ is the FT of the signal/pattern $s(x,y)$. It was only after the filtering capability of masks was recognized that it was realized that the famous Foucault knife-edge test and equally famous Abbé-Porter experiments were actually real (*i.e.* not complex) filtering operations. Indeed the Foucault test is a special case of the *Schlieren method.* The phase-contrast microscope of Zernicke, on the other hand, is a complex filtering operation. More recent applications of these concepts are in pattern recognition and synthetic aperture radar.

4.4.1 Generation of a complex mask

The problem in the preparation of a complex mask is that a complex function must be recorded on a film as variation of intensity which can only be positive and real. With the holography techniques discovered by Gabor and the modulation concepts available in communication theory Leith and Upatnieks solved the problem by generating an offset hologram, the offset chosen so as to separate undesirable frequency components in the output (*see* Chapter 5). A transparency $f(x,y)$ whose FT is to be recorded is placed in the front focal plane of a lens and illuminated by a coherent beam of light (Figure 4.9). A part of the light beam is let to pass on to a prism which refracts it by the desired angle θ. The two together produce an interference pattern on a photographic plate placed at the output plane. The irradiance on the photographic film, $I(u,v)$, is given by

$$I(u, v) = |H(u, v) + A e^{-jk\beta v}|^2 = A^2 + |H|^2 + A H e^{jk\beta v} + A H^* e^{-jk\beta v}$$

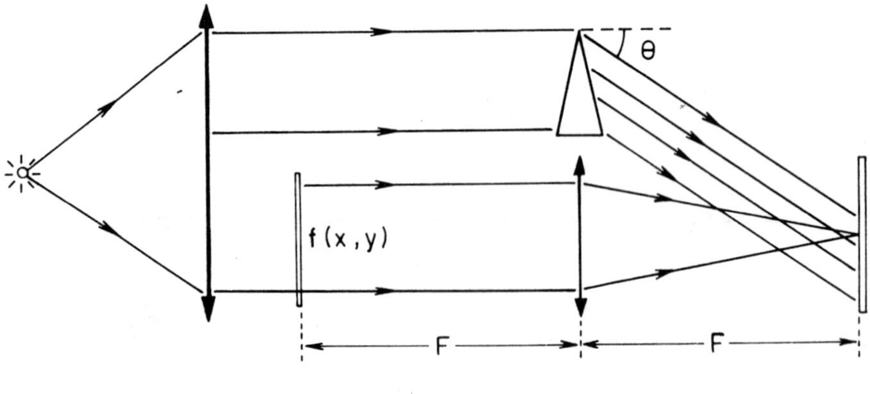

Figure 4.9 Arrangement for generating a frequency-plane mask for a Vander Lugt filter

where $\beta \equiv \sin\theta$. If this mask, after due development *etc*, is placed in the u,v-plane of the signal processor like that shown in Figure 4.8 we get the FT of the product of $I(u,v)$ and $F(u,v)$ giving rise to the output $g(r,s)$ as

$$g(r, s) \propto A^2 f + h \otimes h_-^* \otimes f + h \otimes f \otimes u_0(u, v + \beta) + h_-^* \otimes f \otimes u_0(u, v - \beta) \qquad 4.28$$

where $h_-^* = h^*(-x, -y)$. These four terms are shown in Figure 4.10. If β is so chosen that the terms around the origin can be clearly separated from the sideband terms we can obtain the convolution as well as correlation of f and h.

4.5 Incoherent signal processing

We have already seen in Chapter 3 that the coherence function $\Gamma_g(r_1, s_1, r_2, s_2)$ of the output $g(r,s)$ is given in terms of the coherence function of the input as

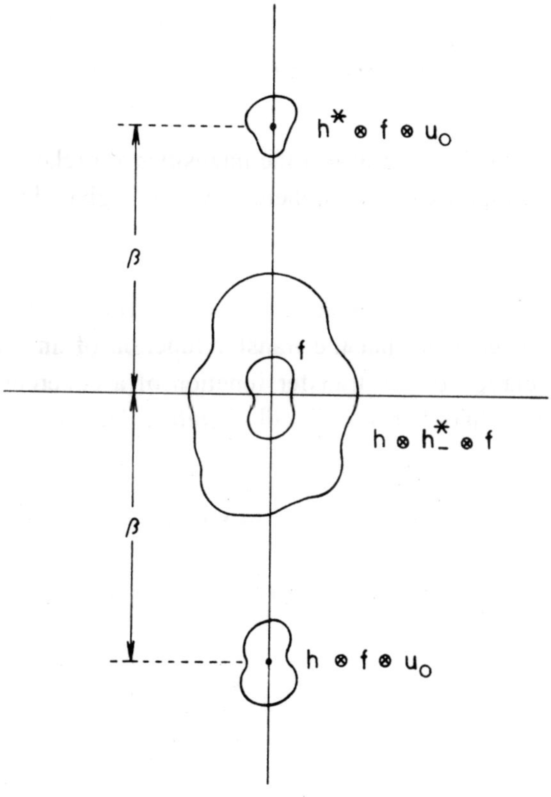

Figure 4.10 Location of the various terms of the output of a Vander Lugt processor

$$\Gamma_g(r_1, s_1, r_2, s_2)$$

$$= \int_{-\infty}^{\infty}\int\int_{-\infty}^{\infty}\int h(r_1 - x_1, s_1 - y_1)\, h^*(r_2 - x_2, s_2 - y_2)\, h\, \Gamma_f(x_1, y_1, x_2, y_2)\, dx_1\, dy_1\, dx_2\, dy_2$$

For incoherent illumination we have

$$\Gamma_f(P_1, P_2) = I(P_1)\; u_0(P_1 - P_2)$$

Substituting into the expression for Γ_g we obtain

$$\Gamma_g(r_1, s_1, r_2, s_2) = \int_{-\infty}^{\infty}\int_{-\infty}^{\infty} h(r_1 - x,\; s_1 - y)\; h^*(r_2 - x, s_2 - y)\, I(x, y)\, dx\, dy$$

The irradiance at any point (r,s) in the output is given by

$$I(r, s) = \Gamma_g(r, s, r, s)$$

$$= \int_{-\infty}^{\infty}\int_{-\infty}^{\infty} |\, h(r - x, s - y)\,|^2\; I(x, y)\; dx\, dy \qquad 4.29$$

Thus in the case of incoherent illumination the intensities are related by a linear transformation and the impulse response of an incoherent system is given by

$$h_{incoherent}(x, y) = |\, h_{coherent}(x, y)\,|^2 \qquad 4.30$$

A consequence of this is that the transfer of any incoherent optical system cannot be of highpass or bandpass type. Note that the transfer function of an incoherent processing system is the autocorrelation of the transfer function of a coherent processing system. Other subtraction methods like "masking" will, therefore, have to be resorted to.

Bibliography

GOODMAN J W, 1968 *Introduction to Fourier Optics* (New York: McGraw-Hill Book Company) Chapters 5,6,7

IIZUKA K, 1987 *Engineering Optics* (Springer-Verlag) Chapters 6,10,11

Chapter 5

HOLOGRAPHY

A S PĀRASNIS

5.1 Introduction

Most of the information about the outside world which man and other beings gather is by the process of "seeing". The process involves the generation of, transmission of, reception of and interaction with light waves. Often there is coding and decoding of the waves at some stages. Considering that, hence, man has always wanted to record, store and retrieve this information it is rather surprising that even after modern science took off the ground it has been only recently that a *true* process of doing all this was discovered. Diffraction photography (*i e* the pinhole camera), lens-image photography and other processes of recording light waves are not "true" in that they record in two dimensions information that is basically three-dimensional; so "truth" is lost. *Holography* is a process that preserves not only the dimension of depth but also parallax. The second characteristic cannot be overemphasized; indeed it is this that makes holography *conformal*, *i e* one that preserves angular and locational relationships among the different elements of a scene. This property is so striking that holography may be said to have triggered a renaissance of optical science.

The history of what is now called holography is not long but it is quite diverse and involves unexpected fields like communication theory, x-ray diffraction, electron microscopy, Fourier theory, *etc* and concepts like standing light waves, spatial frequency, Bragg-reflected light waves, *etc*. The history must begin with Lippmann and Wiener whose work is a hundred years old. Some of the important recent milestones which also give an idea about the uses of holography are given in Table 5.1.

5.2 Nature of light waves

Light waves are electromagnetic. Either of the two expressions

TABLE 5.1

Year(s)	Researcher(s)	Discovery/Invention
1939-42	W L Bragg	X-ray microscope
1948-51	D Gabor	Holography principle and demonstration of recording/reconstruction
1962	E N Leith & J Upatnieks	Off-axis method Use of HeNe laser
1962	Y Denisyuk	Volume holograms Use of HeNe laser
1962	L Mertz & N O Young	Incoherent source holography
1963	Vander Lugt	Matched filtering
1963	Ali Javan & others	HeNe laser (independently)
1964	E N Leith & J Upatnieks	Refinement of laser holography Colour holography
1965	J M Burch Other groups	Holographic interferometry (independently)
1965	A Vander Lugt *et al*	Holographic character recognition
1966	R F Van Ligten & H Osterberg	Holomicroscopy
1967	G W Stroke & R G Zech	Image processing
1967	L D Siebert	Pulsed laser holography
1967	A W Lohmann & D P Paris	Computer-generated holograms
1968	M R J Schwarr	Holographic optical elements
1968	J W Goodman & G Knight O Bryngdahl & W Lohmann	Nonlinear effects
1968	S A Benton	Rainbow hologram
1968	H Nassenstein O Bryngdahl	Evanescent wave holography
1979	J P Huignard	Phase conjugation holography
1982	P Hariharan	Digital holographic interferometry
1986	A S Pārasnis & A Subbārāo	Holography with light caustics

$$\mathbf{E} = \mathbf{E}_0 \cos(\boldsymbol{k} \cdot \boldsymbol{r} - \omega t + \theta)$$

$$\mathbf{E} = \mathbf{E}_0 \exp[j(\boldsymbol{k} \cdot \boldsymbol{r} - \omega t + \theta)] \tag{5.1}$$

for the electric field of a light wave represents a monochromatic plane wave that has the following characteristics, apart from the phase defined by

$$\Phi \equiv \boldsymbol{k} . \boldsymbol{r} - \omega t + \theta :$$

where we have

$\mathbf{E}_0$	—	*polarization vector*
$\mathbf{k}$	—	*propagation* or *wave vector*
ω	—	circular frequency or *pulsation*
θ	—	*phase constant*

We may express the wave vector as

$$\mathbf{k} = k_x \hat{i} + k_y \hat{j} + k_z \hat{k}$$

More familiar are the associated quantities, the wavelength $\lambda = 2\pi/k$, the time frequency $\nu = \omega/2\pi$ and the speed $v = \lambda\nu = \omega/k$. Optical waves, that is what human eyes are sensitive to, have wavelengths in the range of 430-780 nm and frequencies about 10^{15} Hz. The speed of electromagnetic waves in vacuum is a universal constant, $c = 3 \times 10^8$ ms^{-1}. The magnetic field (**B**) associated with the wave of eq 5.1 may be written similarly. **E** and **B** are in phase in empty space and are both normal to **k** such that (**E**,**B**,**k**) is a righthanded set. Eq 5.1 represents a *linearly polarized plane monochromatic electromagnetic wave in vacuum* in the case that the directions of the field remain fixed. Light waves in nature–produced by the sun, lightning, lamps flames, glowworm, *etc*–are far more complicated than this ideally cosinusoidal wave. It is only laser light that comes close to the ideal.

When reference to an actual field encountered in an experiment is made the real part of **E** will be implied. We may also conveniently delete the factor $\exp j(\mathbf{k}.\mathbf{r}-\omega t)$; in any case, it disappears when the irradiance $|\mathbf{E}^2|$ is taken.

5.3 Detectors

All detectors of light (*e g* the human eye, all animal eyes, photographic emulsions, photovoltaic cells, photomultiplier tubes, photodiodes, phototransistors, *etc*) are sensitive to the electric field, not magnetic. Indeed detectors are sensitive to the *irradiance* – formerly called *intensity* – falling on them, that is to $(\mathbf{E}^*.\mathbf{E})$ or $|\mathbf{E}|^2$ rather than **E** itself. Using the trigonometric form of eq 5.1 we have

$$\mathbf{E}^2 = \mathbf{E}_0^2 \cos^2(k \cdot r - \omega t + \theta) \quad 5.2$$

The time variation is so rapid that no known detector can follow it from instant to instant. A detector will give a *time-averaged* reading:

$$I \equiv \langle \mathbf{E}^2 \rangle = \mathbf{E}_0^2 \langle \cos^2 \Phi \rangle = \mathbf{E}_0^2/2 \ \ or\ \ I \equiv \mathbf{E}^* . \mathbf{E} = \mathbf{E}_0^2 \quad 5.3$$

5.4 Plane and spherical waves

For simplicity, but without loss of generality, let us deal with waves travelling in a given, say z-, direction. Let us also assume that the electric fields of all waves have the same direction, so that a scalar description will suffice. Such a wave is represented by

$$E = A\cos(kz - \omega t + \theta) \text{ or } A\exp[j(kz - \omega t + \theta)] \qquad 5.4$$

The wavefronts, *i e* loci of constant phase at a given instant, are then planes normal to the direction Oz of propagation and may schematically be represented as in Figure 5.1 at a given instant. Hence the term *plane wave*. An approximate practical realization is the output of a convex lens at whose focus is a point source.

The constant phase surfaces of a wave originating from a point source at finite distance are concentric spheres. A diverging wave is described by the expression

$$E = (A/r)\cos(kr - \omega t + \theta) \text{ or } (A/r)\exp[j(kr - \omega t + \theta)] \qquad 5.5$$

where r is the distance from the source. A converging wave is described by

$$E = (A/r)\cos(kr + \omega t + \theta) \text{ or } (A/r)\exp[j(kr + \omega t + \theta)] \qquad 5.6$$

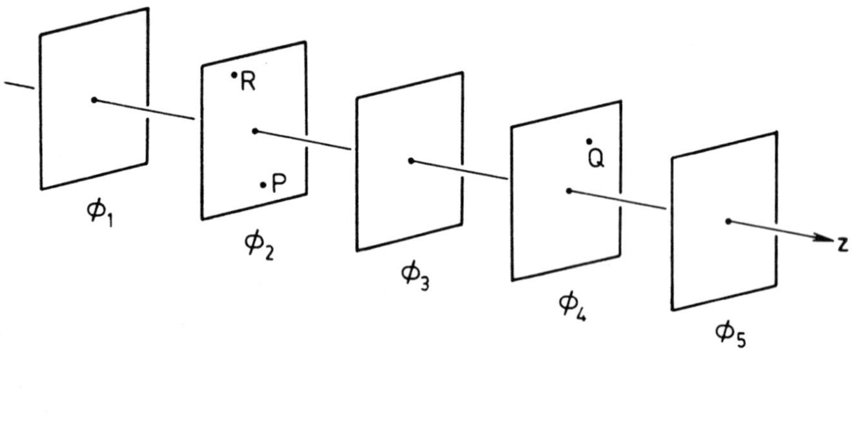

Figure 5.1 Schematic representation of the wavefronts of a plane monochromatic wave travelling parallel to the z-axis

Ideal spherical waves do not exist in nature but an approximate practical realization is the output at the focus of a convex lens if the input is a plane wave. Note that for a spherical wave the magnitude of the wavevector is the same at all points but the direction is along the local **r**. Real waves are quite complicated but can be thought of as the sums of large numbers of different plane or spherical waves.

5.5 Coherence

All phenomena of wave optics (nowadays called *Fourier optics*; the present author has coined the more appropriate term *superposition optics*) are a result of, first, light having a wave character and, second, of the phase relationships between two (or many) arbitrary points considered formally as secondary sources on the same wavefront or on two (or many) wavefronts. Take a point P on the wavefront of phase ϕ_2 and Q on ϕ_4 (Figure 5.1). In the particular case of monochromatic plane waves depicted the phase difference $\phi(Q)-\phi(P)$ is constant for all time even if the actual values of $\phi(Q)$ and $\phi(P)$ vary with time. Since the sinusoidal wave exists throughout infinite space for eternity this is true for any two points on any two different wavefronts although the wavefronts have travelled different distances from the source. The property is described by saying that the wave is *temporally*

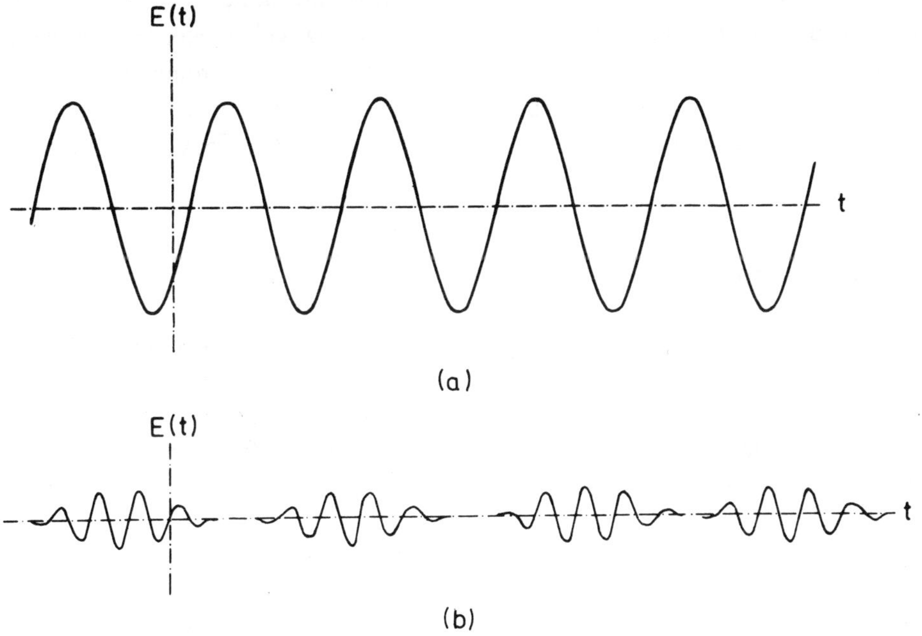

Figure 5.2 The electric field as a function of time of *(a)* an ideal monochromatic wave, *(b)* a *real* light wave

or *longitudinally*, *i. e.* time-of-travelwise, *coherent*. We need to develop the concept of this property because almost always one takes a bit of energy from the regions (of the wavefronts) around the points and mixes or *superposes* them at the detector at some future instant. Temporal coherence is so obvious a property of the monochromatic wave as to be thought trivial.

What if we look at two points P,R on the same wavefront? Would the phase difference be constant? In the case of monochromatic waves the answer is "yes"; in fact the difference is zero. This property is called *spatial* (from "space") or *transverse* coherence.

A monochromatic spherical wave also would possess both properties. Waves from extended sources as we have in practice have limited, or *partial*, coherence of either kind if at all. Coherence can be measured. The time-variation of the electric field of the wave of eq 5.4 is shown in Figure 5.2a (last page). The elementary processes taking place in the atoms which generate photons are, except in lasers, all random so the reality is more like Figure 5.2b wherein each photon is represented by a *wave packet*. Each wave packet may be thought of as being the superposition of a very large number of waves having a range of frequencies and different amplitudes. The successions of randomly generated wave packets, which most actual light waves are, can hardly be called coherent temporally. Wave optical phenomena can still be studied because classical experimental arrangements ensured that random fluctuations would subtract out in turn because two (or more) waves were derived from the same original wave. As to spatial coherence, the arrangement of Figure 5.3 (two holes in an opaque card near a socalled 'monochromatic' lamp such as a sodium lamp, *see* next page) shows that there is no coherence between the waves emanating from S_1,S_2. Nevertheless it should be remarked that *all* sources have some, however small, degree of temporal and spatial coherence.

5.6 Interference

Interference is said to occur when more than one wave exist in a region of space. For it to occur the waves must be at least partially coherent, *i. e.* phase differences must be constant over long enough time. As the most important example let us superpose two equal waves which have travelled different distances from their respective fictitious sources (for example, P,Q or P,R of Figure 5.1) up to the detector, say a plane screen:

$$E_1 = A\cos(kz_1 - \omega t), \quad E_2 = A\cos(kz_2 - \omega t) \qquad 5.7$$

for points P,R and

$$E_1 = A\cos(kz_1 - \omega t), \quad E_2 = A\cos(kz_2 - \omega t + \Delta\Phi) \qquad 5.8$$

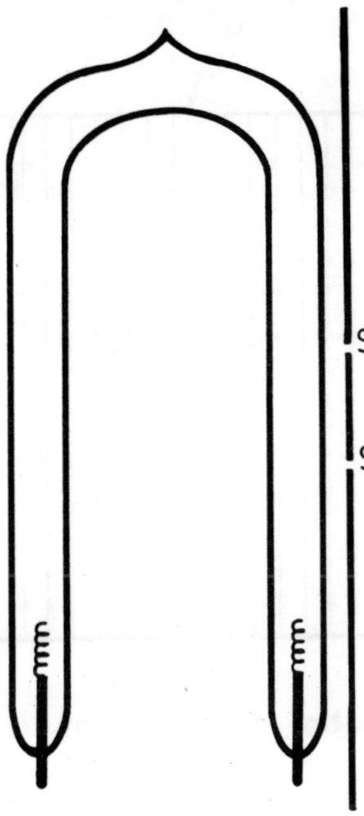

Figure 5.3 Arrangement for studying spatial coherence

for points P,Q – which respectively are examples of interference by *division of wavefront* and *division of amplitude* in practical procedures. Note that they embody the concepts of spatial and temporal coherence respectively. The waves of eq 5.7 having travelled different distances up to the detector, there is a phase difference between them $\delta = k(z_2 - z_1)$. The resultant field is $E = E_1 + E_2$ and the irradiance at the detector, *viz* $|\langle E_1 + E_2 \rangle|^2$, is given by

$$I = I_0 \cos^2(\delta/2) \qquad 5.9$$

δ depends upon the location of the detector. Interference thus consists of an irradiance distribution in space. Where $\delta = p\,2\pi$ we have $I = I_0$, the maxima of irradiance, and where $\delta = (p + \frac{1}{2})\,2\pi$, $I = 0$, the minima. The locus of points at which δ is the same constitutes a *fringe*. We usually talk of *bright* and *dark* fringes for these values of δ respectively.

A fringe is in reality a surface in space. Since, however, most observations in practice are made on plane screens the traces of fringes (historically called 'fringes') are straight or curved lines.

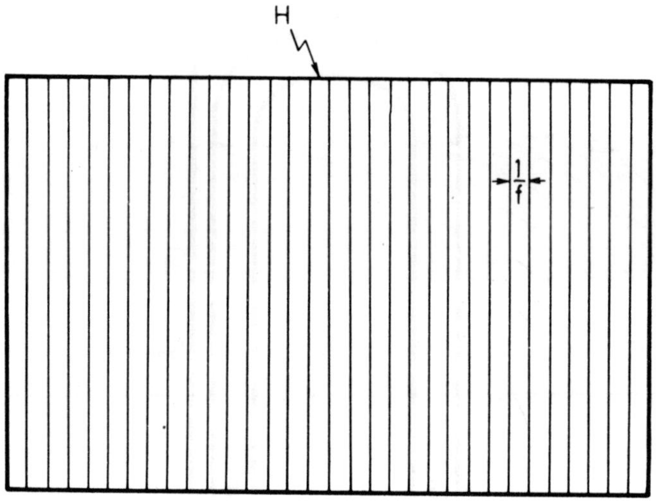

Figure 5.4 Interference fringes of spatial frequency f

As another example, that is most important for our present purposes, consider two plane waves travelling in different directions $\mathbf{k}_1$ and $\mathbf{k}_2$. The maxima of irradiance are obtained where crests meet crests, or troughs meet troughs. By symmetry we see that the space interference pattern is a set of parallel surfaces of maxima and minima normal to the plane ($\mathbf{k}_1$,$\mathbf{k}_2$). If we wish to record this pattern on a sensitive emulsion (H), say a holographic plate, the latter would have to be placed at some angle to that plane; then the emulsion, after processing, will show a series of parallel fringes as in Figure 5.4. The interfringe spacing depends upon the exact geometry. Its reciprocal is called the *spatial frequency* of the fringes. If one of the waves were a spherical one the fringes would be as shown in Figure 5.5 (*see* next page). The fringes in Figure 5.4 look like the lines of a grating but there is one great difference. We define the *transmittance* of a grating (or of the photograph of the fringes) as the fraction of light transmitted by a portion when it is illuminated. The transmittances $t(x)$ of the ordinary grating and the grating of Figure 5.4 are different, as shown in Figure 5.6a,b (*see* p 62). The fringe system of Figure 5.5 looks like, and indeed by calculation may be shown to have identical maxima to those of, a classical Fresnel *zone plate*. The difference again is in their transmittance functions $t\ (r)$ as shown in Figure 6c,d, r being the radial distance from the centre.

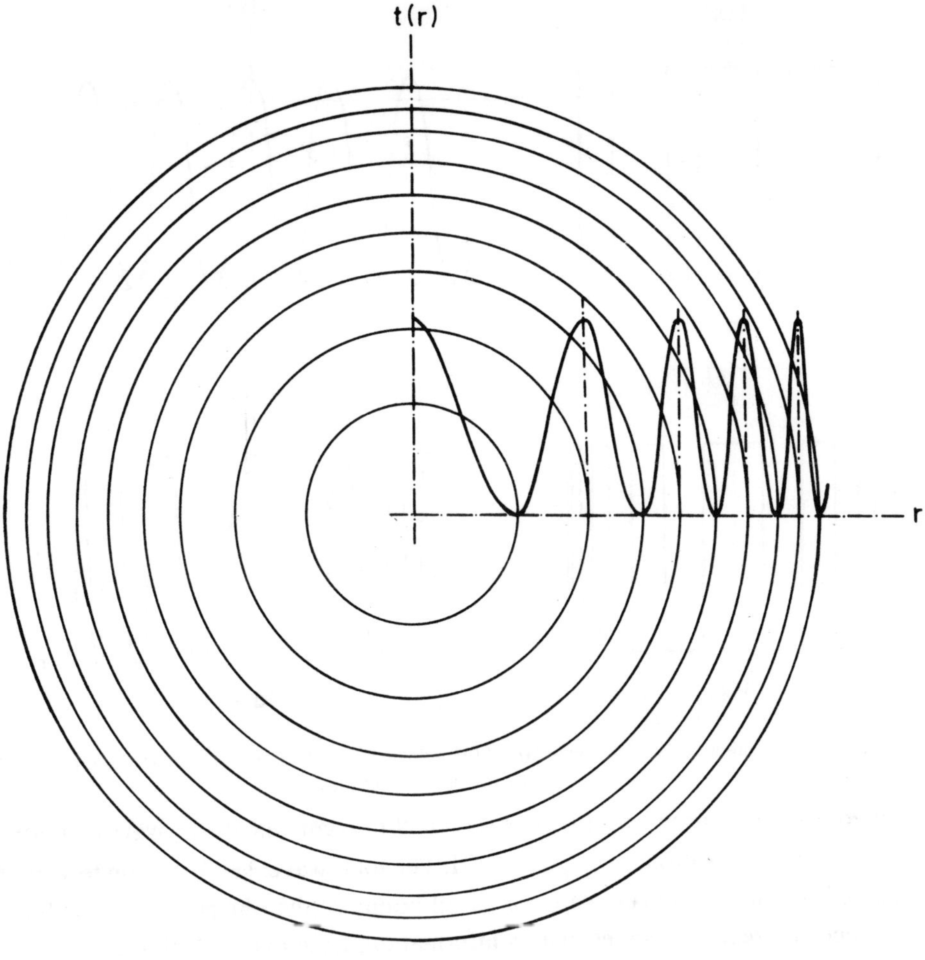

Figure 5.5 Cosinusoidal circular zone plate. The curve shows the variation of transmittance with radial distance

5.7 Holography: Principle and problem

Suppose we want to 'record' an object. In the first place, what does "recording an object" mean? The object has three-dimensionality, size, shape, colour, texture, contours, *etc.* Information about all these must be obtained and stored for retrieval at will. A universal way could be primarily to illuminate the object (O) with light from a source (S) and use a storing detector (H) for the record (Figure 5.7, *see* p 63). The source can be the sun or a lamp. Waves from S, of innumerable frequencies, fall on O; O scatters them; a part of the scattered waves falls on H. The waves falling on H have been modified by the object and

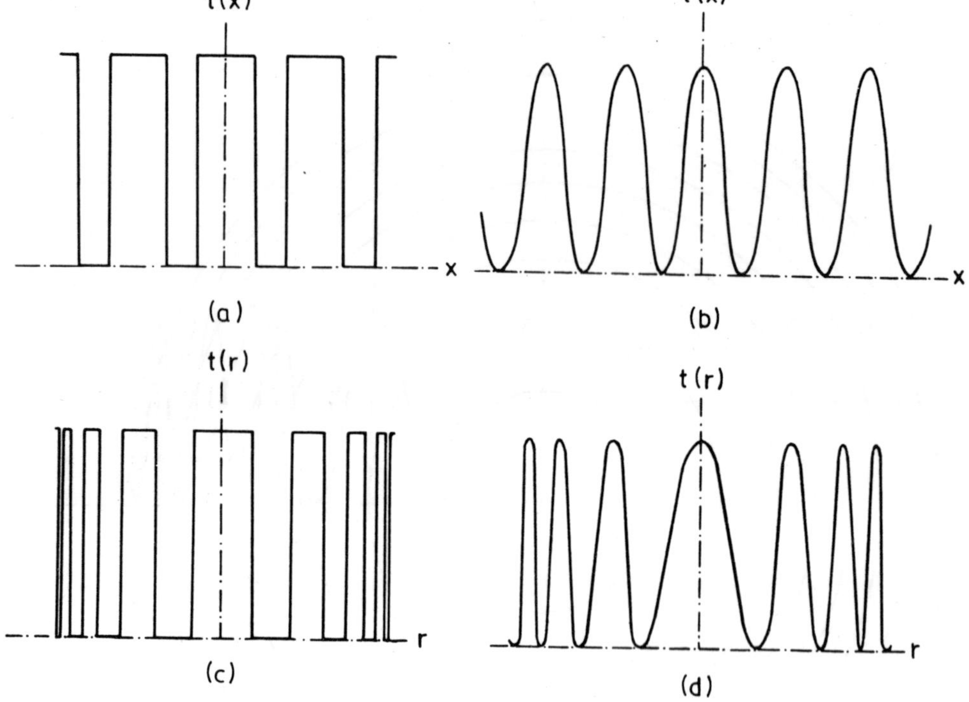

Figure 5.6 Transmittance function $t(x)$ or $t(r)$ of (*a*) ordinary line grating, (*b*) cosinusoidal line grating, (*c*) Fresnel zone plate, (*d*) cosinusoidal circular zone plate

contain detailed information about it. The screen N prevents the direct waves from S, not containing object information, from going to H. The total wave, say Σ_o, going towards H is a complicated mixture of waves scattered by all points of the object. "Each small region (R) of the recorder receives waves, that is information, from every small region (P) of the object and each small region of the object sends waves, that is information, to every small region of the recorder." This is the principle of HOLOGRAPHY in which the word HOLOS means 'whole' and GRAPHIO means 'to write'. 'Writing whole' is what holography is. But we know that the method of Figure 5.7 does *not* in practice yield a record containing information on the object at all! All we get is a fairly uniform irradiance on H. The reason is not far to see. Any one component of Σ_o, say that of frequency ω_i coming from a point P_1 to region R of H may be expressed as $(A/r_1)\exp(j\phi_{1,i})$. H detects and records irradiance, that is the quantity

$$\frac{A^*}{r_1}\exp(-j\,\phi_{1,i}) \times \frac{A}{r_1}\exp(j\,\phi_{1,i}) = \frac{A^*\cdot A}{r_1^2} = \frac{I}{r_1^2}$$

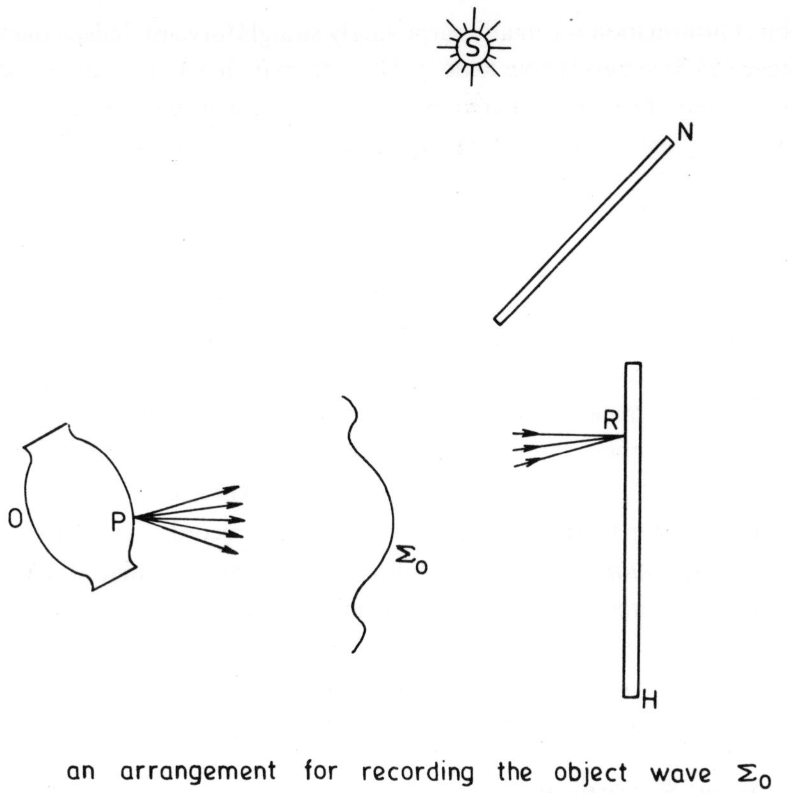

Figure 5.7 Arrangement (that does *not* work) for recording the wave scattered by an object

Hence the information that is mostly contained in $\phi_{1,i}$ is lost. That contained in r_1 is negligibly small. This being true of each frequency component, H has no retrievable information that the object was there at all.

To have a record of some kind, centuries ago man realized that a convex lens placed between O and H would form a record, *viz* the 'image' of O. This method certainly gives a record but it is far from complete. It is a *point-to-point* record so the holographic principle is not satisfied. Further, the rear portions of the object would not be recorded at all. Finally, the record is merely an irradiance distribution in two dimensions of information that is in fact three-dimensional. Quite apart from these shortcomings of a lens it should be stated that the lens performs an amazingly exact function of collecting Σ_o in accordance with the holographic principle but transforming it so that the spherical wave from each 'point' of the object gives rise to an irradiance distribution in the output space that closely resembles the 'point'.

5.8 Holography: Solution

To retain object information is actually surprisingly straightforward. Indeed our discussion on interference (§5.5) suggests how to do it. The information about a small region P of the object is contained in the wave scattered by it towards a particular small region of H. If the phase of the wave scattered by P is θ and amplitude A the expression for the wave reaching R from P is

$$E = \left(\frac{A}{\overline{PR}} \right) \exp[(k\overline{PR} - \omega t + \theta)] \qquad 5.10$$

The problem all along has been to retain what, *viz* $[k(\overline{PR})]$, is contained in the phase; $A/(\overline{PR})$ will always be recorded as $A^2/(\overline{PR})^2$. Figures 5.4 and 5.5 and eq 5.9 suggest that phase information is retained through δ, its difference with the known phase of another wave. Of course, even then, the information manifests itself as irradiance since it is I, proportional to $\cos^2(\delta/2)$, that is recorded by a detector. What better 'another' wave than the source wave itself can there be? And yet while trying to record Σ_o in Figure 5.7 we went out of the way in *preventing* it, by means of the screen N, from reaching H! The recording and its subsequent interpretation would obviously be simple if the source wave were simple, *viz* coherent and plane, and nothing is as coherent and plane as laser radiation; in particular, the light from a helium-neon laser, of wavelength 632.8 nm — the most famous wavelength in the history of modern optics.

5.9 Holography: Recording

The arrangement of Figure 5.8 (*see* p 65) ensures that the object is well illuminated and that the detector-recorder (henceforth to be called the *holographic plate*) receives the object wave Σ_O and the reference wave Σ_R simultaneously. The two waves form a complicated interference pattern on H. After exposure (typically 10^{-2} to 5 s) H is processed photographically as usual, and shows dark and bright fringes interwoven in an intricate pattern. This is the *hologram*, a permanent record of Σ_O. The fringe frequency is so high generally (100-200 mm^{-1}) that to the naked eye the processed hologram appears greyish. Several conditions must be satisfied and precautions taken for the hologram to have high quality, indeed for us to have a hologram at all. These are discussed below.

(*a*) *Vibration Isolation*: Even a laboratory building, leave aside shop or factory floor, contains equipment which gives rise to vibration. A vibration with an amplitude of as little as 10^{-8}m could reduce quality and one with 10^{-7}m obliterate a hologram altogether. Hence the need of isolating the equipment from the floor. All of the components, including the laser, are placed/mounted on the same table top or different table tops appropriately connected. Methods of isolation range from very simple methods such as mounting in/on

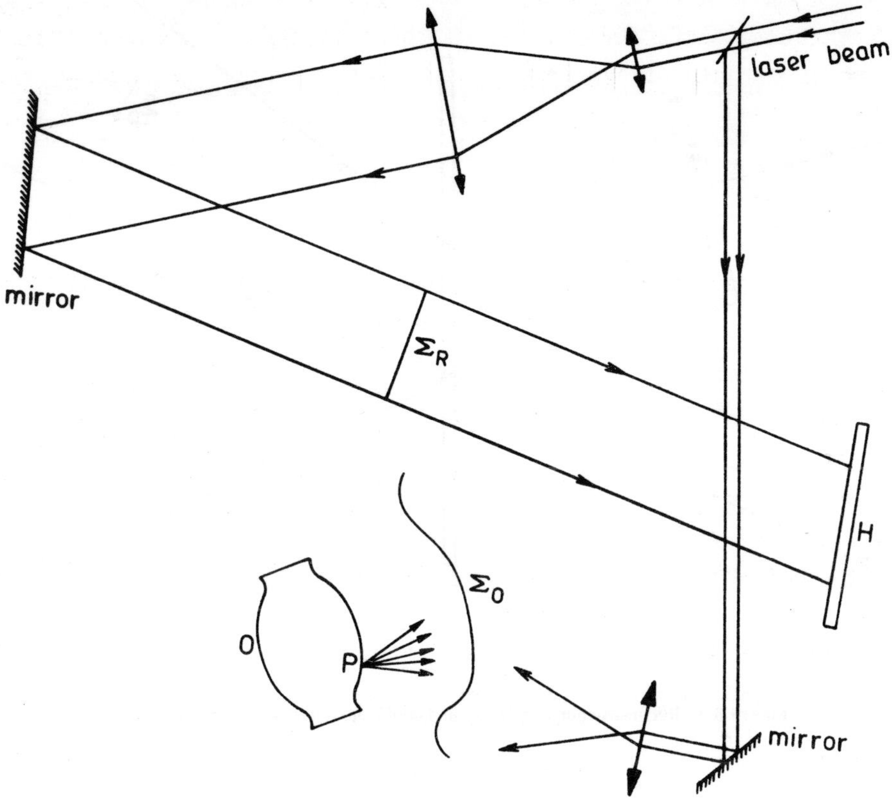

Figure 5.8 Arrangement for recording an off-axis hologram of object O. The double arrowheads represent convex lenses.

fine sand, through 'medium' sophistication-methods — such as placing the table top on inflated vehicle tubes — to highly sophisticated methods like mounting a specially manufactured/structured table top on pneumatic supports which are controlled by servomechanisms. The sandbox costs nothing but is surprisingly effective. Sophisticated systems cost Rs 4 lakh (US $ 10,000) and more (just the table top and supports).

(*b*) *Emulsion Resolution*: Because of the fineness of the fringes the sensitive emulsion of H must have very high resolution, typically 10^3 lines mm^{-1}. That is, two fringes as close as 10^{-3}mm (1 μm) must be seen as separate. Excellent emulsions made by Agfa-Gevaert, Ilford and Kodak are available today. Experimental, laboratory-made emulsions with a resolution of 5×10^3 mm^{-1} have been reported.

(*c*) *Beam Ratio*: Eq 5.9 gives the irradiance distribution when two waves equal in amplitude interfere. The quality of the fringes is measured by their contrast defined as

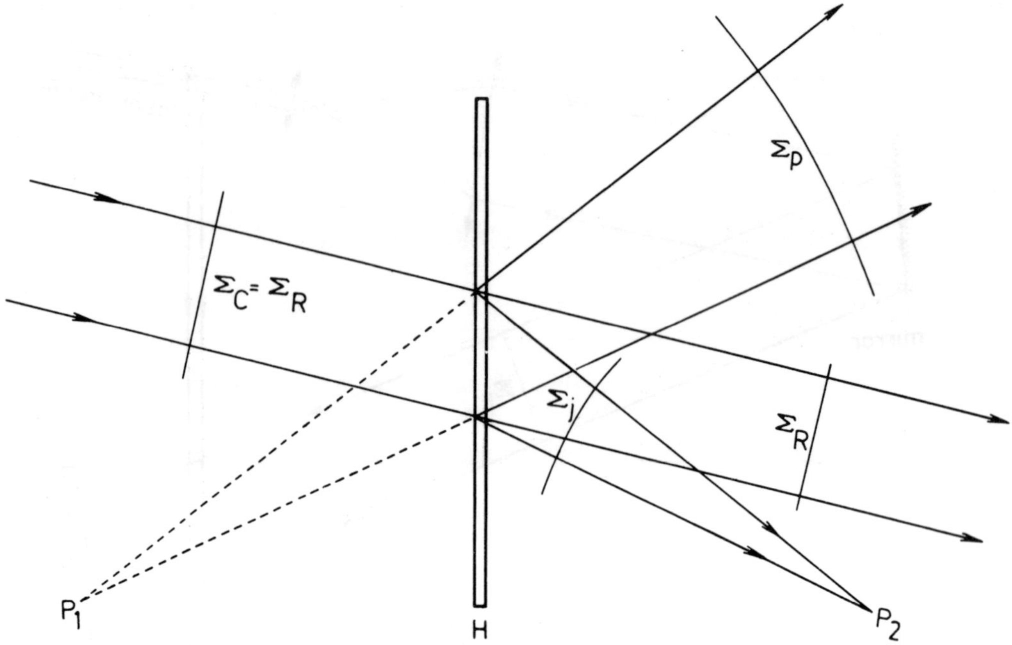

Figure 5.9 Reconstruction of primary and conjugate object waves by a readout wave

$$\gamma = \frac{I_{max} - I_{min}}{I_{max} + I_{min}} \qquad 5.11$$

For two equal waves, clearly $\gamma = 1$. If no interference pattern were formed, that is if the two waves were incoherent, $I_{max} = I_{min}$ and $\gamma = 0$. The contrast thus is *also* a measure of coherence. For coherent reference and object waves of amplitudes A and B respectively, we have $I_{max} = (A + B)^2$, $I_{min} = (A - B)^2$ and

$$\gamma = \frac{(A+B)^2 - (A-B)^2}{(A+B)^2 + (A-B)^2} = \frac{2AB}{A^2 + B^2} \qquad 5.12$$

It is found that a contrast of 0.6-0.8 is ideal. The beam ratio A:B should thus be between 2 and 3. If the beams are equal in strength the contrast varies over the hologram wildly between 0 and 1, with undesirable effects. [*See* (d) below, §5.12 and §5.24]

(*d*) *Exposure*: The holographic emulsion, like all such emulsions including ordinary photographic films, can record within a range of exposure (exposure being measured in $J\ m^{-2}\ s^{-1}$) and, within this range, linearly over a smaller range. The range of irradiation on the holographic plate should therefor be narrow enough so as to avoid distortions and multiple images due to nonlinear behaviour of the emulsion.

(*e*) *Bleaching*: The developed-fixed-washed-dried hologram is quite dark so, when subsequently illuminated, absorbs much light. In order to increase its brilliance (or *efficiency*) it is bleached so as to convert the dark colloidal silver into a transparent salt. One popular method of bleaching is treating the hologram with an acidic dichromate solution. Greater irradiance on any small region of H at the stage of recording gives more silver after normal processing, and thicker/larger salt in that area after bleaching. The distribution of silver dichromate retains the recorded information as phase variation rather than amplitude variation. The details of the reconstruction process are different from those of the unbleached hologram but need not concern us here. The difference between the two kinds of hologram is the same as between ordinary (*i. e.* amplitude) gratings and phase gratings like Michelson's *echelon* grating.

(*f*) *Total Processing*: In view of the high spatial frequencies involved the slightest distortions must be avoided. Processing therefore must be done with care and the control on the temperature of the various baths must be within less than a degree, and dust and debris settling on the hologram avoided.

(*g*) *Paths traversed by the two waves* Σ_O *and* Σ_R: Even a laser beam has limited temporal coherence. It is therefore necessary that the paths traversed by the object and reference waves be nominally equal. This assures *high fidelity* reconstruction.

5.10 Holography: Information retrieval (reconstruction)

The processed hologram, it was said earlier, appears greyish to the naked eye. Under a microscope one would see fringes; in a high quality hologram several different spatial frequencies can be discerned. Neither way does one see anything resembling the object. The object or the object wave needs to be *retrieved* or *reconstructed.* This is done simply by flooding the hologram by a wave Σ_C identical to Σ_R (Figure 5.9). The interference record consisting of silver grains (or silver dichromate grains if bleaching has been done) diffracts light and gives rise to three diffracted waves. One is the direct wave travelling in the direction of Σ_C that contains no object information. Another is a diverging wave which seems essentially to come from behind H. This is called the *primary wave* Σ_P and leads an observer to see the object as it was. We call it the *virtual* image. The third is an essentially converging wave and gives a real image in front of H, this is the *conjugate wave* Σ_J. How exactly do we see the object in this way and is there really a correspondence between Σ_P, Σ_J on the one hand and Σ_O on the other?

5.11 Holography: Elementary analysis

The hologram consists of the interference pattern due to the superposition of Σ_O and Σ_R. Σ_O itself is a superposition of the scattered waves from the different points of the object. Consider one such, P of Figure 5.8. The wave from P, together with Σ_R, would form a set of rings (Figure 5.5), the centre being where the perpendicular from P meets H. If this set alone existed on H and were illuminated by a plane wave Σ_R it would be like illuminating a Fresnel zone plate, with the difference pointed out earlier (Figure 5.6c,d). Such a zone plate focusses a plane wave like a convex lens (at P_2) and, at the same time, acts like a concave lens, with a focus at P_1 (Figure 5.9). Thus we get the exact reconstruction of Σ_O. A classical zone plate (Figure 5.6c) would have given a series of points P_1, P_1' P_1'', ... and P_2, P_2', P_2''... rather than P_1,P_2 only. The object is made of illuminated points like P. Each produces together with Σ_R a zone plate that, on illumination by the readout wave Σ_C, reconstructs it exactly where it was relative to the hologram. The total collection of these points is the object itself. *It must be noted that the retrieval process does not reconstruct the object, nor its image, but it reconstructs the object wave.*

The most remarkable property of the whole process is that it gives a three-dimensional reconstruction. The spatial relationships of the object points are retained. The parallax between any two points is retained. Each illuminated part of the object from where waves reached H while recording is retained in the record. By moving his head an observer could see all such parts exactly as he would while observing the object itself.

Arising out of the basic property emphasized above is the following. If a process merely reconstructed an image of the object, even three-dimensionally, it is not very useful. Holography reconstructs the object wave, a real entity that can, subsequently and at will, be operated upon. It can be made to interfere with another wave, for example. This leads to holographic interferometry (*see* Chapter 6).

Yet another difference from an image formed by a convex lens should be noted. If a photograph is torn into pieces each piece will show only that part of the image that is recorded on it. A piece of a hologram, however, will reconstruct the entire Σ_O because of the hologram relationship (§5.7 above).

There have been in the past two methods of socalled three-dimensional photography: the *stereoscopic* and the *polaroidoscopic*. Both are *pseudoscopic* or false in the sense that, while there is apparent depth to a scene, parallax is not preserved. An object behind another cannot be made visible by moving the head as one would do in reality. Holography, on the other hand, is realistic.

A hologram recorded on an emulsion can obviously not be erased. Nowadays it is possible to employ reusable recording materials: magnetic tape, photochromic materials, thermoplastics, *etc.*

Holographic emulsions have, it may be mentioned in passing, a potential capacity for information storage that, if materialized, will be unmatched. A single square centimetre would have a capacity of about 100 million bits.

5.12 Three output waves

Denoting the mutually coherent object and reference waves by $O(x, y)$, (x, y) being the coordinates in the hologram plane, and R respectively, we have the following expression for the irradiance at H:

$$\begin{aligned} |O(x,y) + R|^2 &= (O^* + R^*)(O + R) \\ &= |O|^2 + |R|^2 + R^* O + RO^* \end{aligned} \qquad 5.13$$

If it falls within the linear range of the emulsion a photographic "positive" of the resulting hologram would have a transmittance given by

$$\begin{aligned} t(x,y) &= t_0 + K|O + R|^2 \\ &= t_0 + K|O|^2 + K|R|^2 + KR^*O + KRO^* \end{aligned}$$

where K is a constant characteristic of the emulsion and t_0 a constant background. Now H is illuminated by a readout wave that is identical to R. The output field of H that goes towards the observer then is given by

$$\begin{aligned} E(x,y) &= R\,t(x,y) \\ &= t_0 R + KR|O|^2 + K|R|^2 R + K|R|^2 O + KR^2 O^* \end{aligned}$$

which, on collecting terms, becomes

$$E(x,y) = [t_0 + K(|O|^2 + |R|^2)]R + (K|R|^2)O + (KR^2)O^* \qquad 5.14$$

The first term is simply proportional to the readout wave R and represents a wave that travels in the same direction and has no object information content. This is the socalled *dc component* of the light diffracted by H. The second term gives the primary reconstruction and the third the conjugate one. To get the irradiance that the observer sees, we take $|E(x, y)|^2$, which will have nine terms. Each could be interpreted separately but the essential point of the arrangement and discussion of Figures 5.8-9 is that, by virtue of obliquity of illumination by R on H, the three terms of eq 5.14, *i. e.* the three output waves, are spatially separated and could be seen independently of one another. This is the *off-axis* method of

Leith and Upatnieks; in Gabor's original linear method they were all in the forward direction and could be observed with difficulty.

5.13 A point object at infinite distance

For a simple but exact analysis of the recording and reconstruction processes and of the three output waves let us suppose that the object is an illuminated point at infinity (Figure 5.10a, *see* next page). The object wave then can be written as

$$O(x) = O_o' \exp(j\, \mathbf{k} . r)$$

$$= O_o' \exp[j(k_x x + k_z z)]$$

$$= O_o \exp[j(k \sin \alpha_0) x]$$

It is superposed with a fully coherent reference wave

$$R(x) = R_0 \exp[j(k \sin \alpha_r) x]$$

on a holographic plate H which is at z = constant. The phase difference at a point of coordinate x is given by

$$\delta(x) \equiv \phi_p - \phi_0 = k(\sin \alpha_r - \sin \alpha_0) x$$

At points where x is such that $\delta = p\, 2\pi$, p being an integer, there is a bright fringe parallel to the y-axis. The interfringe separation is given by

$$\Delta x = \frac{\lambda}{\sin \alpha_r - \sin \alpha_o} \equiv a, \textit{ say}$$

After due photographic processing this would constitute a grating with spatial frequency

$$\frac{1}{\Delta x} = \frac{\sin \alpha_r - \sin \alpha_o}{\lambda}$$

but its transmittance would be cosinusoidal (Figure 5.6b) rather than square wave (Figure 5.6a). When illuminated by $\mathbf{k}_R$ (as in Figure 5.10a) this grating will diffract and give maxima at angles according to the usual grating equation for oblique illumination

$$a(\sin \alpha_r - \sin \theta) = p\lambda$$

which, after solving for θ, gives

$$\sin \theta = \sin \alpha_r - p(\sin \alpha_r - \sin \alpha_o) \qquad 5.15$$

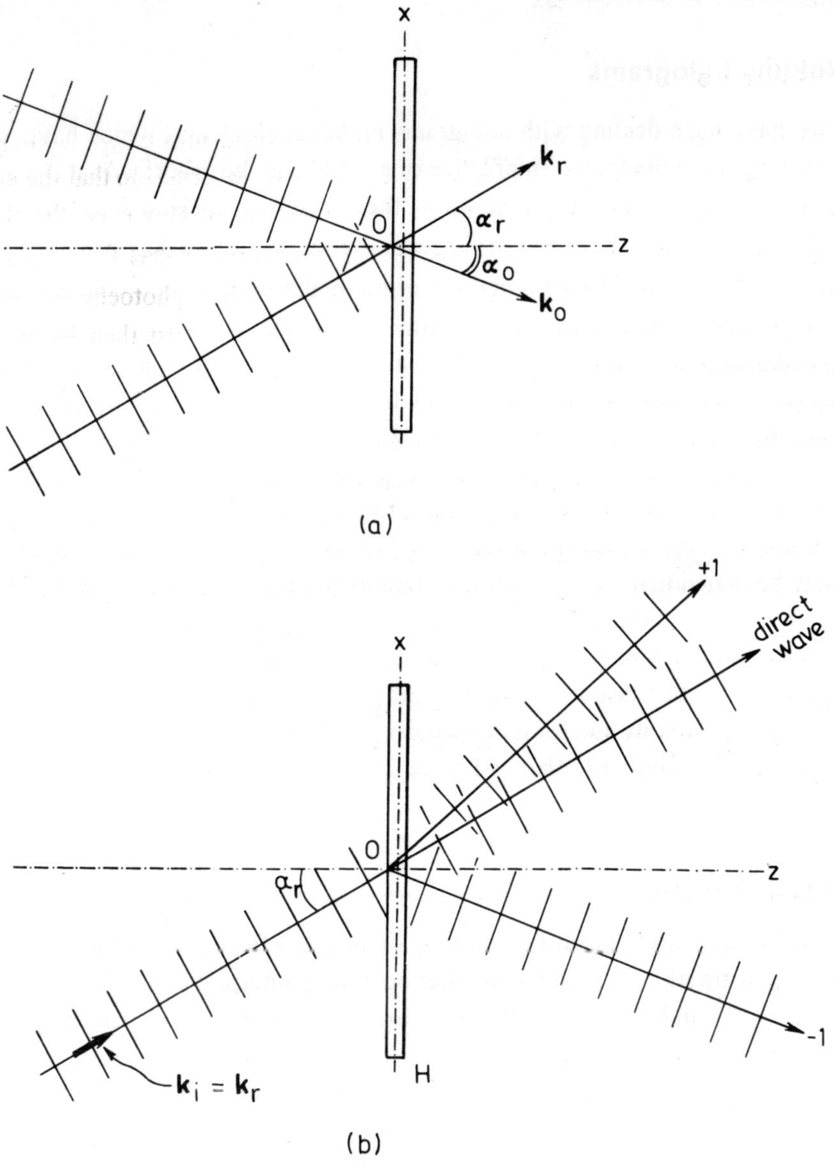

Figure 5.10 (*a*) Recording and (*b*) reconstruction of the plane wave from a point object at infinity

However, since the grating is sinusoidal we need consider only p = 0, ± 1. p = 0 gives the *direct* wave at $\theta = \alpha_1$; +1 gives the *primary* wave at $\theta = \alpha_0$, *i. e.* exactly at the angle at which the object wave had been incident; p = –1 gives the *conjugate* wave at $\theta = \arcsin(\sin\alpha_r -$

sin α_o). The wave at $\theta = \alpha_0$ above is the representation of that coming from a single point object an infinite distance away.

5.14 Volume holograms

So far we have been dealing with holograms made on emulsions which have, mathematically speaking, zero thickness. In practice one could use the criterion that the emulsion be much thinner than the average spacing of the fringes on it. However, the thickness of holographic emulsions (about 20 μm) is generally much greater than fringe spacing which is typically 0.1–1.0 um. That of other recording media such as photochromic and thermoplastic materials is far greater, several mm. The hologram must then be considered as having a distribution of *nodal* (that is, dark) and *antinodal* (that is, bright) surfaces in three dimensions rather than of dark and bright fringes in two dimensions. In the amplitude hologram they are alternately highly-absorbing and less-absorbing layers whereas the phase hologram essentially consists of grains or crystallites of a silver salt embedded in the original gelatinous film on glass or plastic substrate. Or it may be a relief hologram where the thickness through which the waves pass depends on (x,y); in addition, the refractive index may be dependent on x,y,z. All such factors put together only mean that the phase of the readout wave is modulated as it traverses the hologram. The transmittance function in either case of volume hologram therefore needs to be expressed as $t(x,y,z)$ rather than $t(x,y)$. Whereas in the former case it is real, it is imaginary in the latter, and in general complex. For the sake of simplicity and for understanding how image formation occurs we shall only consider a situation similar to that of Figure 5.10. The generalized theory is quite complicated.

5.15 Bragg reflection

Volume holograms are a beautiful example of Bragg reflection, so famous in x-ray/electron/neutron diffraction, for which no other optical counterpart exists. Figure 5.11a shows alternating nodal and antinodal surfaces which form a three-dimensional pattern. Each surface is only partially transmitting. The readout wave is partially reflected from these Bragg planes and a maximum is obtained only if the Bragg equation

$$2\,d\sin\theta = p\,\lambda \qquad 5.16$$

is satisfied. For arbitrary values of d, θ, λ it will not be satisfied but by changing the angle of viewing it can be. The eye will then see a virtual image as shown in Figure 5.11b. A real conjugate image will be obtained if the readout wave is *reverse-reflected* exactly (Figure 5.11c). If white light is used for illuminating, the images will have the colour(s) corresponding to the wavelength(s) which is (are) Bragg-reflected. A slight change in the angle

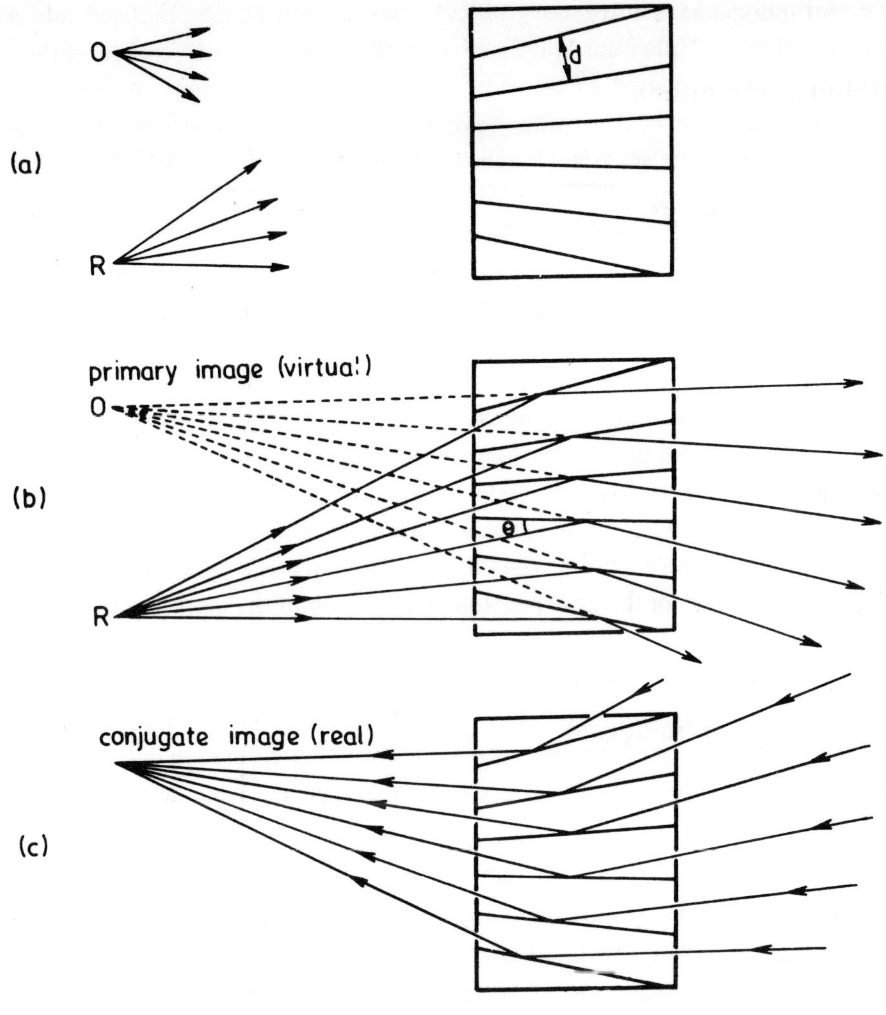

Figure 5.11 Volume hologram of a point object at a finite distance: (*a*) recording, (*b*) reconstruction of primary image (in this case, virtual), (*c*) reconstruction of conjugate image (in this case, real). *d* is the distance between adjoining space fringes.

of illumination will Bragg-reflect another wavelength and the colour of the images will change. As elsewhere so here it must be remembered that the physical process is one of diffraction but it is amenable to a description in terms of (partial) reflection because eq 5.16 holds good at diffraction maxima.

The particular case of Figure 5.10 consists of a point reference source at infinity and a point object, also at infinity but angularly separated. The three-dimensional fringe spacing and inclination are constant. This would not be so for points at finite distance, as already depicted in Figure 5.11. An arbitrary object wave Σ_O can be decomposed into a spatial spectrum of plane or spherical waves so that the resulting interference pattern in the hologram may be considered as the superposition of many fringe patterns of differing spacing and inclination in the plane wave-decomposition, and differently located and scaled ring patterns in the spherical wave-decomposition.

Our discussion of the reconstruction of Σ_O and Σ_O^* by plane or two-dimensional holograms was based on the assumption that diffraction was weak. Even the conceptual explanation of the operation of volume holograms just given is based on this assumption. The assumption is not valid if diffraction efficiency, *i e* the quantity, for example, $(|O^2| + |O^*|^2)/|R|^2$, is high — in other words, the hologram appears brilliant, not dull — because precisely then will the readout wave be strongly attenuated by successive diffraction as it travels through the hologram. At any point in the hologram there are present two waves, one of which, *viz* the diffracted one, gains amplitude at the expense of the other, *viz* the readout wave. This is the basis of what is called the *coupled wave theory*. The Bragg condition is at least approximately satisfied by these two waves. It is worth noting that coupled wave theory is not the only possible explanation of the working of holograms but it is the best one.

5.16 Other kinds of hologram

Reflection holograms are made by having Σ_O and Σ_R arrive at H from opposite sides. The readout wave, after diffraction from the hologram, returns to the same side, so the name *reflection* holograms. Those considered till now were of the *transmission* type. The dependence of the diffraction maximum angle θ on the wavelength is all but eliminated so that if the hologram is seen in white light the brilliant image also appears white. Thus no laser is required for reconstruction. If, while recording, the aperture of the reference wave is limited to, say, the horizontal direction (by putting a long and narrow horizontal slit in the path) then in reconstruction the different wavelengths of the readout wave are separated out. By slightly tilting the hologram an observer could make the reconstructed object waves of different colours (red to violet) go towards his eye at will. This is the socalled *rainbow hologram*.

Except in the case of white light holograms the images have the colour of the laser radiation used: for example, red (632.8 nm) of HeNe, green (543.5 nm) of GreNe — that is, *green*-radiating HeNe laser —, green (514.3 nm) of Ar+, *etc*. In other words, they are *monochrome*. How nice it would be if there are colour holograms! Those who visited the Science Exhibition of the *Festival of USSR in India* in New Delhi or Bombay in 1988 cannot have failed to be wonderstruck at the beautiful reflection colour hologram of differently coloured Soviet medals of honour.

It would be equally fascinating to have holographic movies. Efforts are on way in Hollywood and elsewhere to develop a technique so we can see *real* three-dimensional movies as against the famous *House of Wax* of the middle fifties that was merely pseudoscopic.

In many large metropolises of the world (London, Moscow, New York, Tokyo) there are Museums of Holography. They display unbelievingly realistic objects and scenes of quite complicated kinds. The famous art gallery in St Petersburg, *The Armitage*, has sculpture catalogues that consist of holograms rather than ordinary photographs! The experience of seeing an object realistically long after it has been removed is one that cannot be forgotten easily if ever.

5.17 Noise in holographic images

Any unwanted light falling on the detector produces fringes leading to *noise*. This may result in a reduction in contrast or the appearance of spurious images. There are other causes too. The main causes of noise are as follows:

(*i*) *Granularity*: Conventional silver halide emulsions have, after processing, a granular structure and this leads to a veil around the true image. Thermoplastic materials which are essentially grainless or photochromic materials which have grains of molecular size do not contribute to noise.

(*ii*) *Scattering from the support material or binder*: Optical imperfections of the substrate and support materials, *viz* glass or polyester and gelatine, contribute an inevitable noise.

(*iii*) *Phase noise*: The self-interference of the diffuse object light, which also is inevitable, gets recorded as rather low frequency gratings. At reconstruction they throw light into and around the images with consequent loss of fidelity.

(*iv*) *Nonlinearity noise*: This occurs whenever the dynamic range of exposures falls beyond the linear portion of the detector sensitivity curve. If the range is artificially limited and more energy thus is thrown into the undeviated, that is the *dc*, part of the output the diffraction efficiency suffers.

(*v*) *Speckle Noise*: Speckle noise is intrinsic to the (laser) coherent radiation that is generally used. It is "noise" in the sense that the image is degraded by it. There are several methods of reducing speckle noise, the simplest being stopping down the lens of the imaging system.

5.18 Detailed analysis of Holography

Having built a conceptual basis for holography we shall now analyze the recording and retrieval in a more formal, detailed and exact manner using diffraction theory. Consider an object point O (x_O, y_O, z_O) on an object and a point reference source R (x_R, y_R, z_R) sending their waves F (x_O, y_O) and R (x, y, z) = R_0 [exp (jks)]/s to point P(x, y, z) of H (Figure 5.12). Unless the object is transparent — which we shall not assume here — the point (x_O, y_O, z_O) is any point on a surface that is the object as we see it. Let θ be the angle between the normal to that surface at the point and OP. By eq 2.18 the field at P due to O is given by the Fresnel-Kirchhoff expression

$$O(x, y, z) = -\frac{jk}{4\pi}\int_{-\infty}^{\infty}\int_{-\infty}^{\infty} F(x_o, y_o)\,(1+\cos\theta)\,\frac{e^{jkr}}{r}\,dx_o\,dy_o \qquad 5.17$$

Now r is given by

$$r = \left[(x_o - x)^2 + (y_o - y)^2 + z_o^2\right]^{1/2}$$

which, on expanding, becomes

$$r \approx z_o + \frac{x_o^2 + y_o^2}{2z_o} + \frac{x^2 + y^2}{2z_o} - \frac{xx_o}{z_o} - \frac{yy_o\,(x_o - x)^4}{8z_o^3} - \frac{(y_o - y)^4}{8z_o^3} + \ldots \qquad 5.18$$

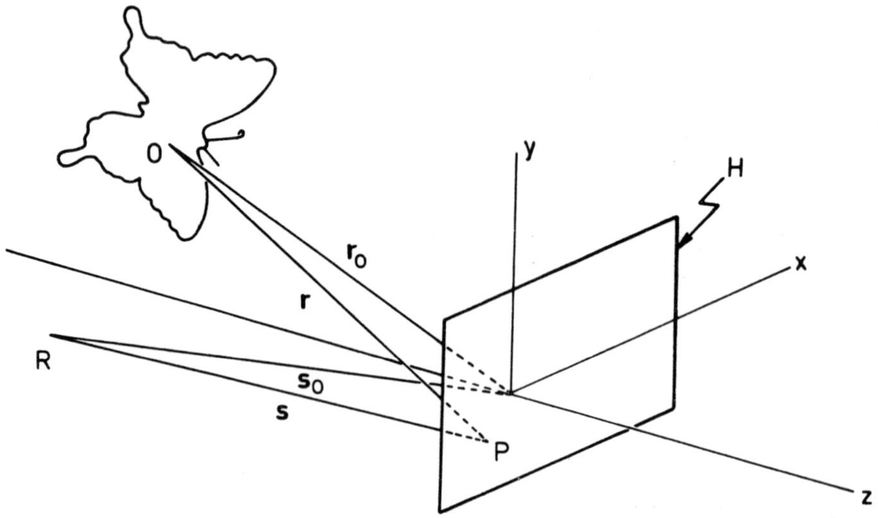

Figure 5.12 Schematic diagram for the calculation of the optical field at point (x,y,z) of a plane hologram

If z_0 is large enough that terms in x_0, y_0 higher than the linear can be neglected, that is, the far-field condition

$$z_0 >> \frac{\left(x_o^2 + y_o^2\right)}{\lambda}$$

is satisfied, and R is plane, H is called a *Fraunhofer hologram*; if not, *Fresnel hologram*. In the latter case usually terms up to and including the quadratic only need be retained.

5.19 Fresnel holograms

Writing the object and reference waves as

$$O(x, y) = O_o(x, y)\, e^{j\phi_o(x, y)}$$

$$R(x, y) = R_o(x, y)\, e^{j\phi_P(x, y)}$$

where O_0 and R_0 are real we have the field at the hologram as

$$H(x, y) = O_o\, e^{j\phi_o} + R_o\, e^{j\phi_P} \tag{5.19}$$

and the eventual transmitted field at the hologram at reconstruction as

$$\left[\frac{1}{K}\right] E(x, y) = C_o\, O_o^{\,2}\, e^{j\phi_z} + C_o\, R_o^{\,2}\, e^{j\phi_z}$$

$$+\ C_o\, O_o\, R_o\, e^{j(\phi_z + \phi_o - \phi_P)} + C_o\, O_o\, R_o\, e^{j(\phi_z - \phi_o + \phi_P)} \tag{5.20}$$

where the readout wave is written as

$$C(x, y) = C_o(x, y)\, e^{j\phi_z(x, y)}$$

and the constant t_0 of eq 5.14 has been dropped for brevity. Eq 5.20 is the fundamental equation of holography. Its salient features are brought out easily by considering a point object (x_0, y_0, z_0) and a spherical reference wave from R only:

$$O(x, y) = B\, \frac{e^{jkr}}{r}$$

$$R(x, y) = A\, \frac{e^{jks}}{s}$$

Let us express the phase at (x, y) of any of the waves O,R,C with respect to that at the origin. For example, after due approximation, we get

$$\phi_O = k(r - r_0) \approx \frac{k}{2 z_O}\left(x^2 + y^2 - 2 x x_0 - 2 y y_0\right) \tag{5.21}$$

which is the form any ϕ (x, y) would have if it were referred to a point source or sink. The exponent of the primary wave term in eq 5.20 may be written, apart from j, as

$$\phi_O - \phi_P + \phi_X = \left(\frac{k}{2\,Z_p}\right)(x^2 + y^2 - 2\,x\,X_p - 2\,y\,Y_p) \qquad 5.22$$

which has the same form as eq 5.21 and wherein the quantities

$$X_p \equiv \frac{x_O\,z_R\,z_C - x_R\,z_O\,z_C + x_C\,z_O\,z_R}{z_R\,z_C - z_O\,z_C + z_O\,z_R}$$

$$Y_p \equiv \frac{y_O\,z_R\,z_C - y_R\,z_O\,z_C + y_C\,z_O\,z_R}{z_R\,z_C - z_O\,z_C + z_O\,z_R} \qquad 5.23$$

$$Z_p \equiv \frac{z_O\,z_R\,z_C}{z_R\,z_C - z_O\,z_C + z_O\,z_R}$$

may be interpreted as the coordinates of the centre of the spherical wave of phase ϕ_π. For a readout wave (C) that is identical to the reference wave (R), *i. e.* C = R, the centre is

$$X_p, Y_p, Z_p = x_O, y_O, z_O$$

which is the object point O exactly! The primary wave thus is an exact representation of the object wave. The centre of the conjugate wavefronts is similarly seen to be

$$X_c \equiv \frac{x_C\,z_O\,z_R - x_O\,z_C\,z_R + x_R\,z_C\,z_O}{z_O\,z_R - z_C\,z_R + z_C\,z_O}$$

$$Y_c \equiv \frac{y_C\,z_O\,z_R - y_O\,z_C\,z_R + y_R\,z_C\,z_O}{z_O\,z_R - z_C\,z_R + z_C\,z_O} \qquad 5.24$$

$$Z_p \equiv \frac{z_O\,z_R\,z_C}{z_O\,z_R - z_C\,z_R + z_C\,z_O}$$

which, for C = R, is

$$X_c \equiv \frac{2\,x_R\,z_O - x_O\,z_R}{2\,z_O - z_R}$$

$$Y_c \equiv \frac{2\,y_R\,z_O - y_O\,z_R}{2\,z_O - z_R} \qquad 5.25$$

$$Z_c \equiv \frac{z_O\,z_R}{2\,z_O - z_R}$$

Now z_O is always negative since the object is on the left side of H. Where the *p*- and *c*-images are, *i. e.* whether they are virtual or real, depends on the signs of Z_p and Z_c respectively, which in turn depend on z_R and z_C. Both images thus could be virtual or real. Notice that for a plane reference wave and an identical readout wave the primary wave reconstructs the object exactly and the conjugate wave forms a real image that is equidistant from H on the positive side.

5.20 Fraunhofer Holograms

Under Fraunhofer conditions the field at (x, y) is given by (*see* Chapter 2)

$$O(x, y) \approx \left(\frac{-jk}{4\pi z_O}\right) e^{jkz_o} \int_{-\infty}^{\infty}\int_{-\infty}^{\infty} F(x_O, y_O)\, e^{-j(k/f)(x x_O + y y_O)}\, dx_O\, dy_O \qquad 5.26$$

where f is the focal length of the lens used to effect Fraunhofer conditions at finite distances (Figure 5.13). It is thus seen that the field at H is approximately the Fourier transform of the transmittance function of the object. In the event that the reference wave is plane H is called a Fourier transform hologram. If $z_O \gg \sqrt{(x^2 + y^2)}$ the lens is not needed. Suppose the object is a single point in the object plane of Figure 5.13; F (x_O, y_O) then is a *delta*-function $\delta(x_O - a, y_O - b)$. The lens will render plane the spherical waves from it. H is then generated by this wave and the plane reference wave. Indeed the hologram of Figure 5.10 is actually

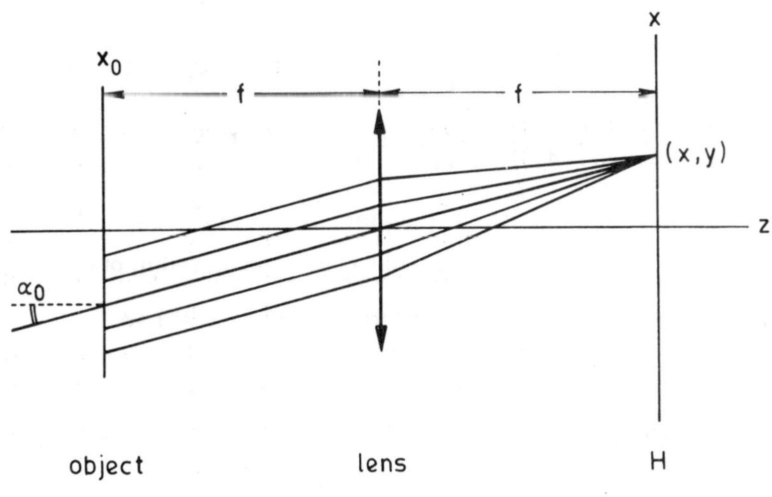

Figure 5.13 Arrangement for recording Fraunhofer, Fourier transform, and lensless FT holograms. (*See* also Figure 5.10.)

an *FT-hologram.* α_O would be given by arcsin(a/f) in the simple case of b = 0. In the basic equation of holography we have $\phi_O = -k\,x\sin\alpha_O$, $\phi_P = k\,x\sin\alpha_R$, $\phi_X = -k\,x\sin\alpha_C$. Again, the object is *exactly* reconstructed if $\alpha_C = \alpha_R$. An FT-hologram may, at least roughly, be obtained by having a 'point' object and reference 'point' source in the same $x_O\,y_O$ – plane at a far distance from H; since the lens, required for creating Fraunhofer conditions, is dispensed with it is called a *lensless FTH.*

5.21 Diffraction efficiency of thin, *i. e.* plane, holograms

A hologram is 'thin' if the thickness of the recording material is small compared with the average spacing of the fringes on it. Its transmittance may be written as

$$t(x,y) = |t(x,y)|\exp[-j\phi(x,y)], \quad 0 \le |t(x,y)| \le 1 \tag{5.27}$$

where the factor before the exponential is the amplitude transmission coefficient at the small element $(dx\,dy)$ at a point (x, y) and $\phi(x, y)$ is the phase retardation due to the element.

(*A*) *Amplitude hologram* In an amplitude hologram $\phi(x, y)$ is essentially constant whereas $|t(x, y)|$ varies over the surface. Consider the situation of §5.13, Figure 5.10. The transmittance of the resulting hologram will be given by

$$|t(x)| = t_0 + t_1\cos[k(\sin\alpha_R - \sin\alpha_O)x] \equiv t_0 + t_1\cos(Kx) \tag{5.28}$$

where t_1 is the amplitude of the spatial variation of $|t(x)|$. The average values of t_0 and t_1 are $\frac{1}{2}$ each so we write

$$\begin{aligned} |t(x)| &= \frac{1}{2} + \frac{1}{2}\cos Kx \\ &= \frac{1}{2} + \frac{1}{4}\exp(jKx) + \frac{1}{4}\exp(-jKx) \end{aligned} \tag{5.29}$$

The diffracted irradiance of either image, therefore, is $1/16$ of the readout incident wave. The diffraction efficiency thus is $1/16 = 0.0625$. This is never achieved in practice.

(*B*) *Phase hologram* If there is no absorption, $|t(x, y)| = 1$ so

$$t(x) = e^{-j\phi(x)} \tag{5.30}$$

We assume that the phase retardation when the readout wave passes through the element $(dx\,dy)$ is proportional to the irradiance there while recording; then

$$\phi(x) = \phi_0 + \phi_1\cos(Kx) \tag{5.31}$$

and the complex amplitude transmittance will be

$$t(x) = e^{-j\phi_0} e^{-j\phi_1 \cos(Kx)} \quad 5.32$$

Neglecting the first factor that is constant, t(x) may be expressed as a Fourier series:

$$t(x) = \sum_{n=-\infty}^{\infty} j^n J_n(\phi_1) e^{jnKx}$$

where J_n is Bessel function of the first kind of order n. The diffraction pattern will consist of a large number of orders, the amplitude in the nth order being proportional to the value of $J_n(\phi_1)$. The amplitude diffracted into the first order, which alone contributes to the image, has a maximum at a value of ϕ_1 a little over $\frac{1}{2}\pi$. Thus we have

$$I_{max} = \left[J_1^{\,2}(\phi_1) \right]_{max} = 0.339 \quad 5.33$$

which is much greater than what obtains for amplitude holograms.

5.22 Fringes in three dimensions

Following upon the concepts of § § 5.14-15 we shall, for brevity, retain the term fringes for the distribution of the nodal and antinodal surfaces in the volume of the recorder. Figure 5.14 gives the geometry of recording. U_R, U_O are the recording fields; $\mathbf{k}_R$, $\mathbf{k}_O$, their wavevectors; Oz is normal to the hologram plane and Ox is in that plane. The angles θ_R, θ_O are in the emulsion; those in air could be found by using the Snell law. A simple geometrical calculation gives the uniform x-spatial frequency of fringes (that is, the reciprocal of the spacing along the x-direction) as

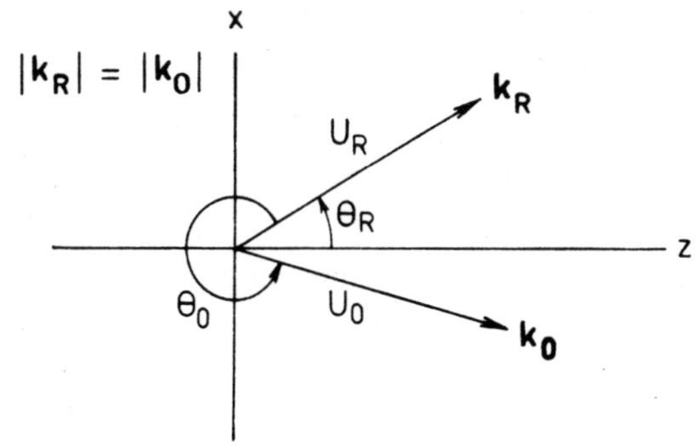

Figure 5.14 Wavevector geometry for recording a volume hologram

$$\frac{\sin\theta_R - \sin\theta_O}{\lambda} \tag{5.34}$$

which is the same as we had in §5.13. The radian spatial frequency would be 2π times this, that is

$$\omega = k'(\sin\theta_R - \sin\theta_O) \tag{5.35}$$

where k′ is the wavevector magnitude in the recording medium. If we define ϕ as

$$\phi \equiv 2\pi - \theta_O + \theta_R$$

this becomes

$$\omega = 2k'\sin\left(\frac{\phi}{2}\right)\cos\left(\theta_R - \frac{\phi}{2}\right) \tag{5.36}$$

The angle made by the fringes with the *z*-axis is given by

$$\alpha = \theta_R - \phi/2 \tag{5.37}$$

So eq 5.36 becomes a neater expression

$$\omega = 2k'\sin(\phi/2)\cos\alpha \tag{5.38}$$

5.23 Three-dimensional grating equation and Bragg conditions

For avoiding needless geometrical complication in the calculation of how this fringe system diffracts the readout wave we shall take the source to be a line rather than a point one (L of Figure 2.3) and suppose that it is parallel to the *y*-axis. This gives a cylindrical symmetry around Oy so that from the Fresnel-Kirchhoff expression for Fraunhofer diffraction (eq 2.32) we may delete that coordinate. Now our diffracting aperture is the hologram, *i. e.* what was the screen in Chapter 4. Making suitable changes for the symbols in the expression for the Fraunhofer diffraction integral we have the field diffracted by an element *dx* of the hologram aperture to the point P_i as

$$U_{P_i} = C\int_H e^{-jk(m_i - m_c)x}\,dx \tag{5.39}$$

Referring to Figure 5.15a,b (on p. 83) for the meanings of the symbols, we have used this notation:

$$l_c = z_c/s_0,\quad m_c = -x_c/s_0$$

$$l_i = z_i/r_0,\quad m_i = -x_i/r_0 \tag{5.40}$$

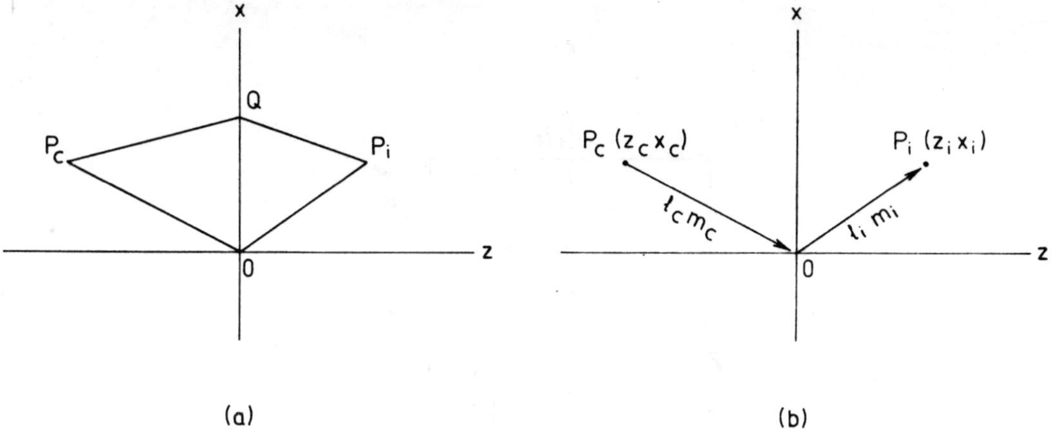

Figure 5.15 Reconstruction by diffraction at a volume hologram

We have put P_c in Figure 5.15 for L of Figure 2.3, and Q for M. At the same time we shall write $U(m_i)$ rather than $U(P_i)$ so as to emphasize the diffraction *direction*. So far the whole description is as we had in the case of the plane aperture (*mutatis mutandis*, as to notation) but here we have a *thick* aperture.

Eq 5.39 gives the diffracted field in the direction m_i, the diffraction being due to an elementary aperture of thickness dz located at $z = 0$. It is implicit herein that the wave diffracted into a direction m_i by any element dz is weak so that each element may, as a first approximation, be considered to receive from its preceding element a wave of identical amplitude. This constitutes the *weak diffraction* assumption. The contribution by an element dz located at $z = z$ to the field in the direction (l_i, m_i) is

$$d[\,U l_i\,, m_i)\,] = \left[\; C \int_{-H}^{H} e^{-jk(m_i - m_c)x}\; e^{-jk(l_i - l_c)z}\; dx \;\right] dz \qquad 5.41$$

The total contribution would be obtained by integrating over z from $-\,t/2$ to $+\,t/2$, t being the thickness of the recording emulsion (Figure 5.16 on p. 84). The hologram-aperture is not a *clear* one; there is a variation in transmittance within dz itself for all values of z. This spatial variation is denoted by $G(x,z)$, the socalled *pupil function*. Then we have

$$U(l_i, m_i) \;=\; C \int_{-t/2}^{t/2} \int_{-H}^{H} G(x, z)\, e^{-jk(m_i - m_c)x}\; e^{-jk(l_i - l_c)z}\; dx\, dz \qquad 5.42$$

The next step, which we shall not take here, is to calculate $G(x,z)$ and substitute into eq 5.42. The result of subsequent integration is a somewhat cumbersome expression that

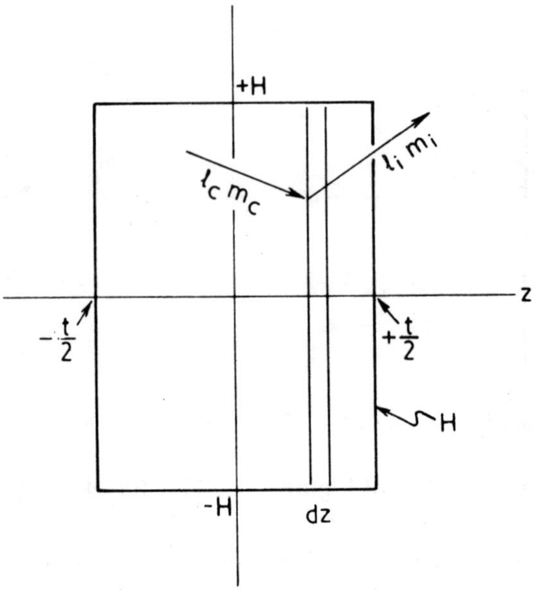

Figure 5.16 Calculation of the contribution to the diffracted field by an element dz of the thickness of a volume hologram

contains three terms. One of them is maximum when $l_i = l_c$ and $m_i = m_c$ which conditions imply that the diffracted wave is in the forward direction, so it is the direct wave containing no object information. The second term is maximum when the following conditions are satisfied:

$$l_i = l_c - \omega_0 m / k$$

$$m_i = m_c + \omega_0 l / k \qquad 5.43$$

where l, m are the direction cosines of the trace of the fringe plane in the zx-plane (Figure 5.17 on p. 85). The third term is maximum when

$$l_i = l_c + \omega_0 m / k$$

$$m_i = m_c - \omega_0 l / k \qquad 5.44$$

In eq 5.43-44 ω_0 is the radian spatial frequency of the three-dimensional fringes. Let us write eq 5.43-44 in a more convenient form as follows:

$$\frac{-(l_i - l_c)}{m} = \frac{\omega_0}{k}, \quad \frac{m_i - m_c}{l} = \frac{\omega_0}{k} \qquad 5.45$$

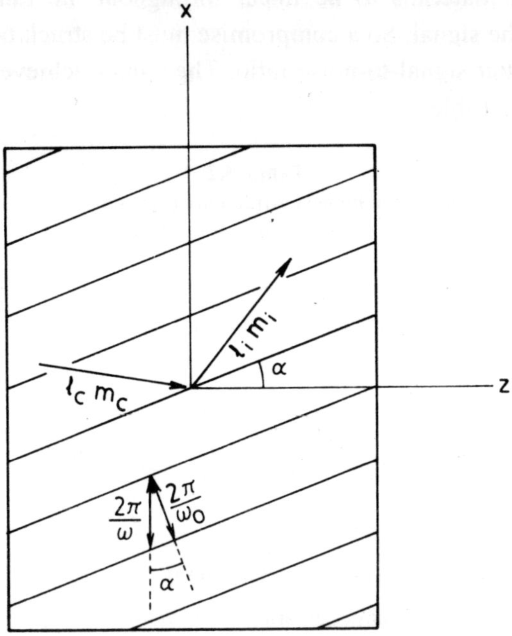

Figure 5.17 Maximum of diffracted field and the Bragg equation

$$\frac{l_i - l_c}{m} = \frac{\omega_0}{k}, \quad \frac{-(m_i - m_c)}{l} = \frac{\omega_0}{k} \qquad 5.46$$

Clearly these directions are opposite of each other. Eq 5.45-46 constitute the Bragg conditions for the primary virtual image and the conjugate real image respectively. In more familiar terms we already had eq 5.16 — the Bragg equation.

5.24 Coupled wave theory

The calculation of diffraction efficiency values in §5.21 for plane holograms was on the assumption that diffraction was weak. Even the explanation of the operation of volume holograms was based on this assumption. The assumption is not valid if diffraction efficiency values are high, because precisely then will the readout wave be strongly attenuated. At any point in the hologram there are present two coherent waves, one of which, *viz* the diffracted one, gains amplitude at the expense of the other one, *viz* the readout wave. This is, as said in §5.15, the basis of the *coupled wave theory*. The Bragg condition is at least approximately satisfied by these two waves; no other orders satisfy them so they are not present. Diffraction efficiency values can then be calculated for various kinds of holograms. Table 5.2 gives the theoretical highest values. In practice the inability of recording materials to be linear throughout the range of exposures adds nonlinearity noise to the signal. So a compromise must be struck between greater diffraction efficiency and better signal-to-noise ratio. The values achieved in practice are about 50% of those quoted in Table 5.2.

TABLE 5.2
Theoretical highest difraction efficiency values
A represents *amplitude*, P *phase*

Kind of hologram	Transmission				Reflection	
Type of hologram	Thin		Thick		Thick	
Modulation	A	P	A	P	A	P
Value	0.063	0.339	0.037	1.000	0.072	1.000

Bibliography

FRANÇON M, 1974 *Holography* (New York: Academic Press)

HARIHARAN P, 1985 *Optical Holography* (Cambridge: University Press)

PĀRASNIS A S, 1990 *Basic principles of holography*, Physics Education (India) **7** (3) 181-193

ROBILLARD J and Caulfield H J, 1990 *Industrial Applications of Holography* (Oxford: University Press)

SMITH H M, 1975 *Principles of Holography* (New York: Wiley-Interscience)

SYMS R R A, 1990 *Practical Volume Holography* (Oxford: University Press)

Chapter 6

HOLOGRAPHIC INTERFEROMETRY

B M SIVARĀM

6.1 Introduction

Holographic interferometry is the technique of using classical interferometry and combining it with the technique of recording wavefronts from three-dimensional diffusively reflecting objects with nonplanar surfaces in order to measure small optical path differences. The technique was first developed by Powell and Stetson in 1965 and is briefly explained below. Figure 6.1 (*see* next page) is a schematic drawing of a holographic setup. The object wave is recorded on the holographic plate as described in detail in the last chapter. After due processing it is replaced in the plate holder without any changes in the rest of the setup.

One can then directly compare the holographic reconstruction of the object against the original scene under identical conditions of illumination. Thus if an opaque card were to be placed in the path of the reference beam one could observe the object through the hologram. If the card were to be placed to block the scene beam one could observe the reconstruction. It is usually dimmer but appears identical to the original scene. By placing a neutral density filter in the scene beam the intensities of the two beams could be matched. If now one person randomly blocks either the scene or the reference beam another person watching through the hologram cannot in fact tell the image from the real object.

If the beams are allowed to pass through the hologram simultaneously one sees the two identical superimposed scenes. When the two are put into near-perfect coincidence by carefully repositioning the hologram the dual image will suddenly be seen to be covered by fringes, alternately bright and dark. Their frequency and orientation can be controlled by the tilts of the hologram. One is now observing an interference between the two beams of light. One beam is generated by the action of the reconstructing beam on the hologram. The other beam is the object wave itself in *real time*. The fringes are due to phase differences between the two beams coming from different pairs of corresponding regions of the two wavefronts. If the superimposition is absolutely exact — which in practice it can never be

— there will be no fringes. By carefully repositioning the hologram it is possible to reduce the number of fringes to one or two in a 10cm × 15cm hologram.

Referring again to Figure 6.1 we see that the fringes also appear whenever the scene is displaced by microscopic amounts provided, of course, that the hologram has been re-mounted to within wavelengths of its original position. Even blowing air through the scene is effective in producing fringes. If the surface of an object is touched gently with, say, a pencil point concentric fringes are produced which map the deformation of the surface over its original state. Each fringe spacing now denotes a change in the altitude of the surface, relative to the viewer, of one-half of the wavelength of the light used. For HeNe radiation this would be about 300 nm or 13 millionths of an inch. If the object is warmed fringes would appear due to microscopic inhomogeneous thermal deformations. If the object were loaded with a small weight the fringes would be due to changes in the state of the surface consequent upon internal stresses. The accuracy is the same as in classical interferometry except that the technique can be applied to non-optical surfaces and situations. One of the discoverers of holographic interferometry, Dr J Burch, discovered it in the following way. He noticed that fringes were seen through the hologram whenever he leaned on the table carrying holographic equipment. He had the wisdom to realize the significance of the effect.

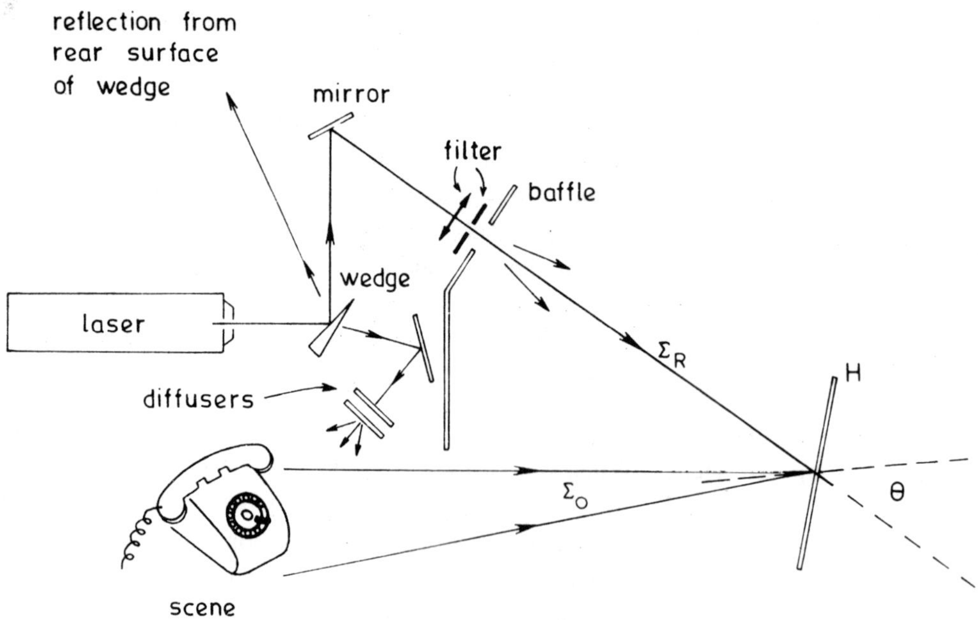

Figure 6.1 Configuration for recording transmission holograms

This technique in which the developed holographic plate is replaced in its holder and the interference between the holographically reconstructed original object wave that was recorded in the past, on the one hand, and the new changed object wave that exists now, on the other, is called *stored beam* or *real time* holographic interferometry. The wavefront against which optical interferometric measurements are made is stored in the hologram and the fringes are viewed at the instant they occur. A variation is to vibrate the subject harmonically and then synchronously modulate the output of the laser. This is called *stroboscopic stored beam* or *stroboscopic real time* holography.

Holographic interferometry can be done more simply by exposing the holographic plate twice: once, before changes have occurred or been made in/on the object and the second time, afterwards. Then small differences which have occurred between the exposures result in a reconstruction of the fringes as well as both of the object waves. The two holograms are locked permanently in the same emulsion and a 3D interference phenomenon is seen whenever the combined hologram is illuminated for reconstruction. This technique was discovered, again by accident, by Wuerker and his colleagues at TRW when a Q-switched laser accidentally double-pulsed. They took a hologram of a flying bullet and found more than its three-dimensional image. There were fringes in the shock wave, as if it had interfered with another wave. It turned out that by accident the laser used had flashed twice, once when the bullet arrived and once after it had arrived and proceeded a little further. The two optical waves scattered from the shock wave were frozen on the hologram one after another with the same reference beam. The readout beam revives them simultaneously and the reconstructed waves interfere with each other. This is the *double exposure* or *elapsed time* method. A third method is *dynamic time average* holographic interferometry. In this a recording is made on a single hologram of the object undergoing cyclic vibratory motions. The interference that occurs at readout is within the entire ensemble of images, with the images recorded near zero velocity contributing most strongly to the holographic exposure.

6.2 Mathematical description

6.2.1 Double exposure method

This method consists in recording two successive holographic exposures of an object in two different positions. When the hologram is reconstructed it produces an interference pattern which is indicative of the object displacement between the two exposures. The mathematical description of double exposure holography is best setup by first considering the recording of two object wavefront fields U_{O1} and U_{O2} in the same emulsion using a reference wave U_R. Since these wavefronts are all additive in the plane of the emulsion the resulting irradiance distribution I recorded in the hologram may be expressed as

$$I = |U_{O1}|^2 + |U_{O2}|^2 + 2|U_R|^2 + U_R^*(U_{O1} + U_{O2}) + (U_{O1}^* + U_{O2}^*)U_R \quad 6.1$$

where the asterisk indicates the complex conjugate. If the plate is processed as usual and illuminated with U_R the reconstructed wavefront will have a term of the form $(U_{O1} + U_{O2})$ which is proportional to the image wavefront U_i:

$$U_i \propto (U_{O1} + U_{O2}) \quad 6.2$$

The reconstructed image described by eq 6.2 is then the vector sum of the corresponding object fields. Since the deformations and displacements considered are generally small compared to the spatial extent of the object the individual object fields can be represented as

$$U_{O1} = A_O e^{jk\varphi_{O1}} \text{ and } U_{O2} = A_O e^{jk\varphi_{O2}} \quad 6.3$$

The irradiance distribution I_{id}, the subscript denoting the image wavefront in the double exposure case, in the reconstructed image is then

$$I_{id} = |U_i|^2 \propto 2|A_O|^2 [1 + \cos k(\phi_{O2} - \phi_{O1}] \quad 6.4$$

wherein $\phi_{O2} - \phi_{O1}$ is the direct measure of the additional path length incurred by the object illumination beam as a result of the object deformation or displacement. The irradiance distribution of the reconstructed image is a function of the irradiance at the object point $|A_O|^2$ modulated by the cosine factor which is a function of the phase change resulting from the object displacement. In order somewhat to simplify the derivation it is assumed that the object displacement z_I is perpendicular to the surface. If now the angle of the incident illumination, as measured from the normal to the object surface in the construction process, is α and the angle of observation, as measured from the normal to the object in the reconstruction process, is β the image irradiance can be written

$$I_{id} \propto 2|A_O|^2 [1 + \cos k z_1 (\cos\alpha + \cos\beta)] \quad 6.5$$

Using the trigonometric identity $\cos^2(x/2) = ½(1 + \cos x)$ this may be expressed as

$$I_{id} \propto 4|A_O|^2 \cos^2 \frac{1}{2} [k z_1 (\cos\alpha + \cos\beta)] \quad 6.6$$

Eq 6.6 describes the geometry of the interference fringes found on the surface of the image as reconstructed from a double exposure hologram. The irradiance varies as the cosine squared, with the distance between successive fringes directly related to the displacement z_I. To summarize, in double exposure holographic interferometry the two recorded states of the object reconstruct as though both had physically existed simultaneously and the interference pattern which results is identical to a pattern obtained from a conventional interferometer.

6.2.2 Time-average method

This technique consists in making a single holographic recording of an object while it is subjected to a cyclic vibratory motion and assumes that the exposure time in recording the hologram is long compared to one period of the vibration cycle. In this manner the hologram effectively stores an ensemble of images corresponding to the time average of all positions of the object during its vibration. Consequently, upon reconstruction, the interference among the ensemble of images, or the vector sum of the individual wavefronts, produces an interference pattern which is weighted toward the deformation extremes of the object, *e. g.* a test piece.

If the vibration frequency is ω and the amplitude z_2 then the object light field can be represented as

$$U_O = A_O \, e^{j k y z_2 \cos \omega t} \qquad 6.7$$

where y is a function of the incident illumination and observation angles. This equation expresses the form of the object field for any instant of time and, since the hologram records every position of the object over one period of the vibratory cycle (T), it is necessary to take the time-average of this equation over one period of the vibratory cycle to obtain the correct form of the image which will be reconstructed:

$$U_{ia} \propto \frac{1}{T} \int_0^T U_O \, dt$$

$$= \frac{A_O}{T} \int_0^T e^{j k y z_2 \cos \omega t} \, dt \qquad 6.8$$

wherein the subscript "*ia*" denotes the image wavefront in the time-average case. The corresponding irradiance in the image is given by

$$I_{ia} = | U_{ia} |^2 \propto | A_O |^2 \left[\frac{1}{T} \int_0^T e^{j k y z_2 \cos \omega t} \, dt \right]^2 \qquad 6.9$$

This equation can be further simplified, knowing that the integral representation of the Bessel function $J_n(x)$ is of the form

$$J_n(x) = \frac{j^{-n}}{2\pi} \int_0^{2\pi} e^{jx\cos t} e^{jnt}\, dt \qquad 6.10$$

For the case n = 0 eq 6.9 may be written

$$I_{ia} \propto |A_O|^2 J_0^2 [kyz_2] \qquad 6.11$$

where J_0 is the zero order Bessel function.

If the angles of illumination (α) and observation (β) are defined as they were for eq 6.6 the irradiance distribution may be expressed as

$$I_{ia} \propto |A_O|^2 J_0^2 [k(\cos\alpha + \cos\beta) z_2] \qquad 6.12$$

It can be seen that eq 6.12 is similar to eq 6.6 except that the irradiance now varies as the square of the Bessel function. This variation is characterized by being maximum when the argument of the Bessel function is zero and by having succeeding maxima which decrease in value. A plot of this function is presented in Figure 6.2. It can be seen from this plot and eq 6.12 that when the displacement is zero the reconstructed image irradiance is the brightest. Therefore regions having no motion or the nodes in the vibratory pattern exhibit the greatest irradiance and can readily be detected upon inspection of the reconstructed image.

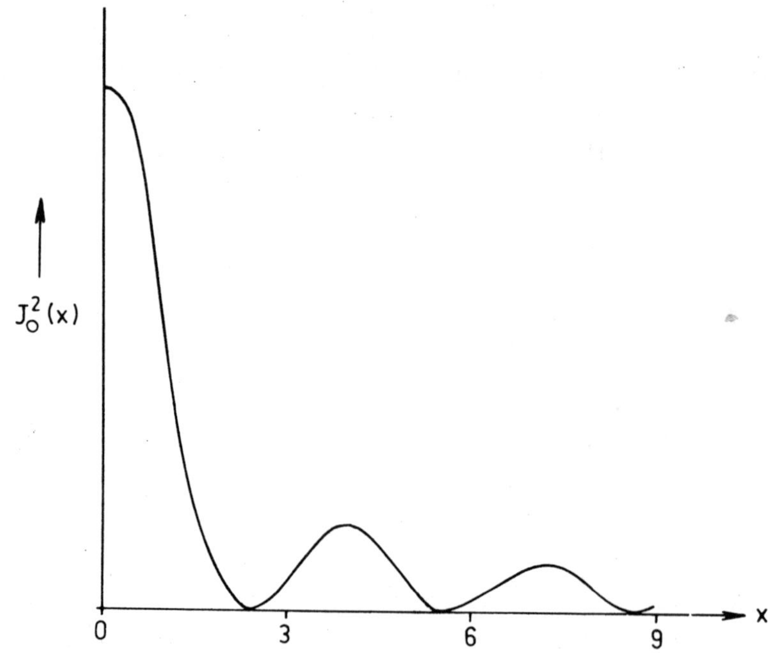

Figure 6.2 Interferometric fringe irradiance variation along a beam as a result of cyclic vibratory motion

In summary, it can be seen that time-average holographic interferometry is a unique tool for studying sinusoidal vibration patterns using only one holographic exposure. The technique provides information about the amplitude of vibration and the location of the vibratory nodes, and is applicable to nonsinusoidal motions also.

6.2.3 Real time method

This technique consists in taking a single holographic exposure of an object in an unstressed state, processing the plate and replacing it in exactly the same position in which it was recorded. (The precision with which the hologram must be replaced is of the order of wavelengths. Therefore in practice the plate is usually processed *in situ* using a variety of developing tanks especially designed for this application.) The hologram is then reconstructed using the identical coherent reference beam used in the construction process. The reconstructed image is now superimposed on to the original object which is also illuminated with the same light as when the hologram was recorded. Interference fringes are seen if the object has been changed slightly. This interferometric comparison between the original state of the object — *viz* the reconstructed holographic virtual image — and the new state of the object is made at the instant it occurs; hence the term *real time*. The particular advantage that this technique has is that different types of motion, dynamic as well as static, of the object could be studied by making a single holographic exposure rather than having to make an interferometric hologram for each.

The theory of real time fringe formation for the general case, where the object undergoes both a static displacement and harmonic vibration, can be derived once the instantaneous irradiance distribution is found and then time-averaged over the integration time of the human eye.

As in eq 6.2 the reconstructed static image wavefront of the original object which was recorded on the hologram is

$$U_i \propto U_O = A_O \qquad 6.15$$

The instantaneous complex disturbance at the object may be expressed as the vector sum of the reconstructed image wavefront and the wavefront now being scattered by the object, or

$$U_{ir} \propto A_O + A_O\, e^{j k y (z_1 + z_2 \cos \omega t)} \qquad 6.16$$

where z_1 represents a static displacement of the real object, independent of time, and $z_2 \cos \omega t$ the harmonic motion of the real object. (The subscript "ir" denotes the image wavefront in the real time case.) Accordingly, the instantaneous irradiance in the object-image space is given by

$$I_{ir} \propto |A_0|^2 \left[2 + e^{jky(z_1 + z_2 \cos\omega t)} + e^{-jky(z_1 + z_2 \cos\omega t)} \right] \qquad 6.17$$

The time-average of this irradiance over the integration time T is derived in a manner similar to that employed for eq 6.12 and is

$$I_{ir} \propto 2|A_O|^2 \{ 1 + \cos[k z_1 (\cos\alpha + \cos\beta)]$$

$$\times J_0 [k z_2 (\cos\alpha + \cos\beta)]\} \qquad 6.18$$

The fringe system in real time holographic interferometry can be described, for the case of pure static displacement ($z_2 = 0$), as

$$I_{ir} \propto 2|A_O|^2 \{ 1 + \cos[k z_1 (\cos\alpha + \cos\beta)]\}$$

$$\propto 4|A_O|^2 \cos^2 \frac{1}{2} [k z_1 (\cos\alpha + \cos\beta)] \qquad 6.19$$

This equation is identical to eq 6.6 for the double exposure case and therefore has the same fringe characteristics. In the case of a harmonically vibrating object observed through its own hologram ($z_l = 0$) the observed irradiance is

$$I_{ir} \propto 2|A_O|^2 \{ 1 + J_0 [k z_2 (\cos\alpha + \cos\beta)]\} \qquad 6.20$$

This equation is not identical to eq 6.12 of the time-average case in that an additional unity term is present and the Bessel function appears in the first rather than the second power. These two factors cause a reduction in the fringe contrast observed in real time holography as opposed to the time-average case. An added difference, resulting from the first power of the bessel function, is that the total number of fringes is half of that found for the time-average case.

In practice the real time observation of harmonic motion in the presence of a static displacement will cause a noticeable decrease in fringe contrast as a result of the term

$$\cos[k z_1 (\cos\alpha + \cos\beta)] \qquad 6.21$$

unless the static displacement z_l is adjusted to be equal to zero.

In summary, real time holographic interferometry offers a technique for studying a variety of motions with a single holographic interferogram but has the drawback that both fringe contrast and sensitivity are reduced when studying vibratory motion. The requirement of film processing *in situ* is a further disadvantage. However, in many situations these drawbacks are far outweighed by the considerable time saving which the method affords.

6.3 Holographic nondestructive testing

Holographic interferometry can be used very effectively for the detection of flaws and dimensional deviations without resorting to destructive techniques of inspection. The information required for making a decision is all contained in the shapes and locations of the interference fringes obtained in the holographic interferogram. This interference portrait directly provides information concerning various internal defects of engineered or industrial products. This technique can be used to delineate regions of reduced strength in shells and outer casings of pneumatic and hydraulic devices, defects in bonding of layers and piles in multilayer structures, structural defects such as cracks and fatigue zones in vital components like turbine blades, pressure vessels, *etc*.

The object to be tested is recorded holographically before as well as after it is stressed or loaded. The double exposure hologram will reveal the defects through anomalies in the fringe patterns. These anomalies make it possible to locate the defects and assess their seriousness. The optimal type and method of stressing depends on the specimen being investigated and stress can be applied mechanically, thermally, acoustically, *etc*. The anomaly must be as clear as possible whether the interference pattern is from a double exposure hologram or a real time one. The comparison of the interference portraits of the defective and reference specimens must be facilitated as far as possible by providing such optimal loading conditions that the normal patterns have the simplest possible appearance. The comparison being made and fringe evaluation done by computerizing are now routine procedures.

6.4 Otological studies using holography

6.4.1 Arrangement of apparatus

Holographic otological studies have been carried out on man as well as other animals. The tympanic membrane, the middle ear and the inner ear have been investigated. A schematic diagram of the apparatus used is shown in Figure 6.3 (*see* next page). The apparatus is built around a standard speculum that can be inserted into the outer ear. The beam splitter serves to couple the laser beam into the ear and to couple out the scattered light on to the holographic plate through the optical window W. The reference beam and the scattered light interfere on the holographic plate to give a hologram of the ear structures such as the tympanic membrane. By exciting the loudspeaker at various frequencies and powers time-average holographic studies could be made on the tympanic membrane. Generally the laser energy is kept below 1 mJ cm^{-2} which is the accepted safety level. This setup can also be used for double exposure holography by changing the pressure stepwise inside the closed system instead of pulsating it.

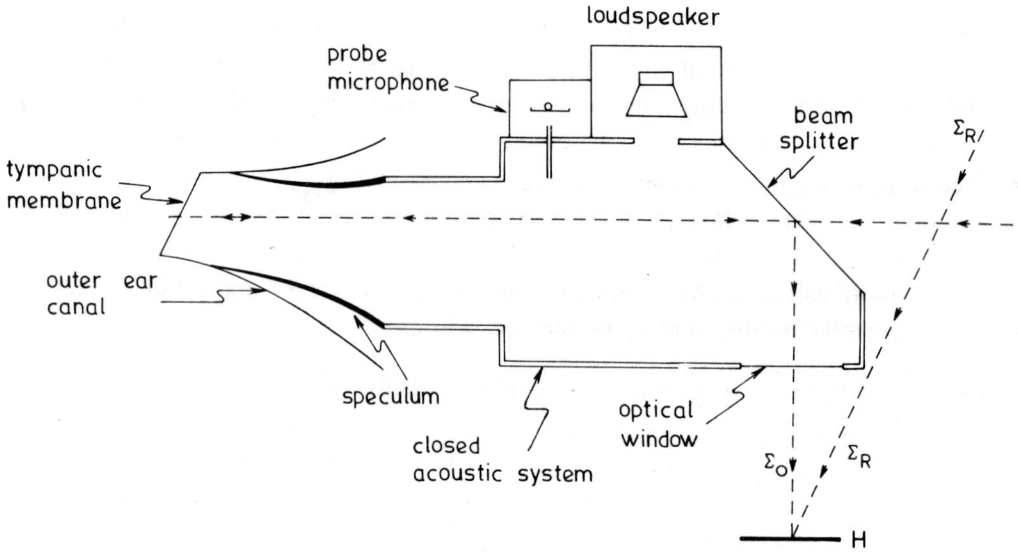

Figure 6.3 Closed acoustic system for sound application and holographic recording of tympanic membrane vibrations through the outer ear canal

6.4.2 Schematic diagram of the ear

Figure 6.4 (*see* next page) shows the structure of the outer, middle and inner ear schematically. The tympanic membrane forms the boundary to the middle ear from the outer ear. The middle ear is located inside the tympanic cavity and contains the ossicular chain, malleus, incus and stapes, and ligaments and muscles. The inner ear is shaped like a coil and the cochlea is separated into the parallel canals by a membrane — called the *basilar* membrane — and the receptor cells for hearing are arranged on it in a special distribution. We shall here not go into the medical details of the findings, the diagram of Figure 6.4 serving only to emphasize that holography alone is capable of such investigations on the miniature structures.

6.5 Bioholography

Holography is not confined to Physics and engineering alone but is wide ranging in its conceptual applicability. Life is based on order and pattern and this conforms well with the fundamental pillars of holography, *viz* coherence and interference. The holography principle is apparently in operation in nature, and bats and cetaceans use such a procedure to navigate and manoeuvre using acoustic signals. The socalled *echolocation* is not based on the classical radar principle whereby ultrasonic pulses are emitted and the echoes from

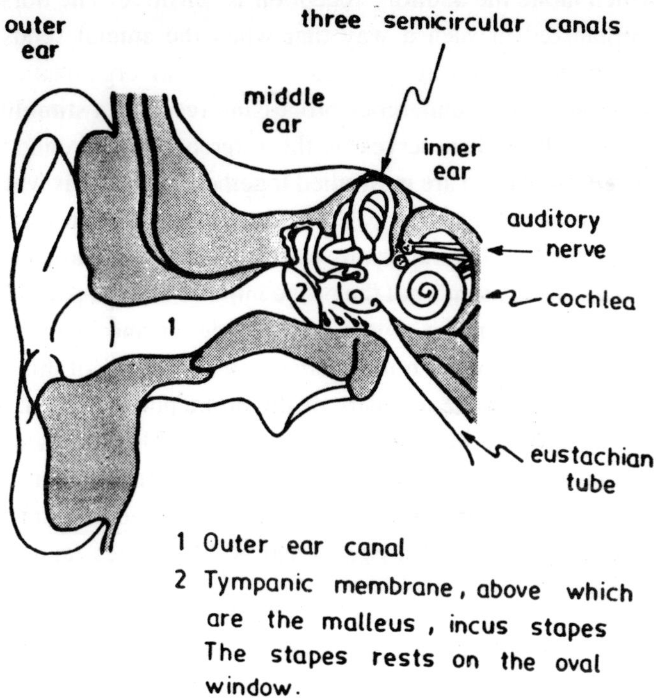

Figure 6.4 Schematic cross section of the human peripheral organ

surrounding objects get picked up. If this were so most bats would just starve even if they escaped collision. The duration of an ultrasonic pulse emitted by a bat is of the order of several tens of microseconds. If one uses the appropriate velocity of sound in the medium then it can easily be seen that reflected pulses from objects closer than 1.5 m will arrive at such an object and after reflection reach the bat even before the outgoing probe pulse has terminated. This means that if the bat used a neural mechanism for crosscorrelating the outgoing probe pulse and the returning echo no target identification would be achieved, since analyzing the envelope of the crosscorrelation function does not give any information on the shape of the object and on whether it is an obstacle to be avoided or a prey not to be. Further, how then do bats distinguish their own cry from cries originating from fellow bats in an environment in which the signal received may be weaker than the noise by a factor of about 2000? The answer lies in a bat's remarkable auditory substitute for 3D vision.

6.5.1 These animals emit coherent waves to get information about their environment. Professor Greguss predicted that the nature of the mechanism that extracts arrival time-information from echoes reflected and translates it into spatial dimensions is based on the holographic principle. The time intervals, *i. e.* the phase-bound information, are transformed *via* a coherent reference background in a reversible way into proportional amplitude relations to which alone the auditory reception is sensitive. The dorsal colliculus of the bat has to be organized in such a way that when the animal sends a stimulus — information — to generate an ultrasonic pulse, the neural organization of the dorsal colliculus where the echo pattern undergoes processing receives a stimulus replica of the outgoing sound pattern which then serves as the reference background. This reference background and the returning echo are multiplied together and it is this which serves as the information-bearing term for processing by the brain. This prediction was verified experimentally by the Russian scientist Ayrapetyans who identified a neural connection between the part of the bat's brain which sends out the probe impulse and the part of the brain which receives the reflected echo from the object and decodes it and analyzes it. He further proved that whenever the background noise increases the intensity of this stimulus pattern increases too without increasing the intensity of the probe pulse emitted. This manifestation of the holographic process in the bat is consistent with the obvious fact that, by increasing the intensity of the coherent background R, the information-carrying term is amplified by 2R. In the light of this it is possible to understand how information can be processed by bats even when the processed signal is 2000 times fainter than the background noise.

Similar holographic processing occurs in dolphins for their navigation.

Bibliography

ERF H K, 1974 *Holographic Nondestructive Testing* (London: Academic Press)

OSTROVSKY Y I *et al*, 1980 *Interferometry by Holography* (Berlin: Springer-Verlag) *Acoustics and Vibration progress Vol 11*, 1976 (London: Chapman and Hall), *Proc Int Workshop on Holography in Medicine and Biology* 1979 (Berlin: Springer-Verlag)

VEST C M, 1979 *Holographic Interferometry* (New York: John Wiley and Sons)

Chapter 7

HOLOGRAPHIC OPTICAL ELEMENTS

P C MÉHTA

7.1 Introduction

The imaging properties of holograms are receiving increasing attention due to their ability of providing unique system configurations. A new class of optical elements, *Holographic Optical Elements or HOEs*, is fast evolving. Their technological aspects as well as design features have recently been studied extensively. Their real potential is not yet fully visualized and awaits further research. Unimaginable tasks are possible by employing HOEs. A plane wavefront gets phase-transformed on passing through a hologram and produces a three-dimensional image — an impossible task for conventional glass optics. A large-sized holographic lens recorded on a film base can be rolled into a small piece that can be unrolled when required. Large optical apertures, at once light-weight and of lower cost, are *plus* points of HOEs. Though considerable theoretical work has been carried out no recognized general designs or ready made successful designs of HOEs are available as yet unlike conventional glass optics where a new design can be based on previous experience. Holographic designers are therefore finding difficulty even in getting a suitable starting point. Besides this the technological difficulties are also yet to be overcome.

7.2 What is an HOE?

A holographic optical element is a hologram of a point source; in other words, it is a photographic recording of an interference pattern between a spherical wave and a plane wave (Figure 7.1, *see* next page), or between two spherical waves. In general, it can be considered as a generalized grating structure. A holographic lens can converge a collimated beam coming from a point source situated at the focus of a convex lens. HOEs can be made to work as gratings, lenses, aspherics or any other type of optical components. HOEs have the following unique characteristics and capabilities:

(*i*) HOEs are diffractive elements unlike glass lenses which are reflective or refractive (Figure 7.2a, *see* p. 101). HOEs are, therefore, wavelength-sensitive. The

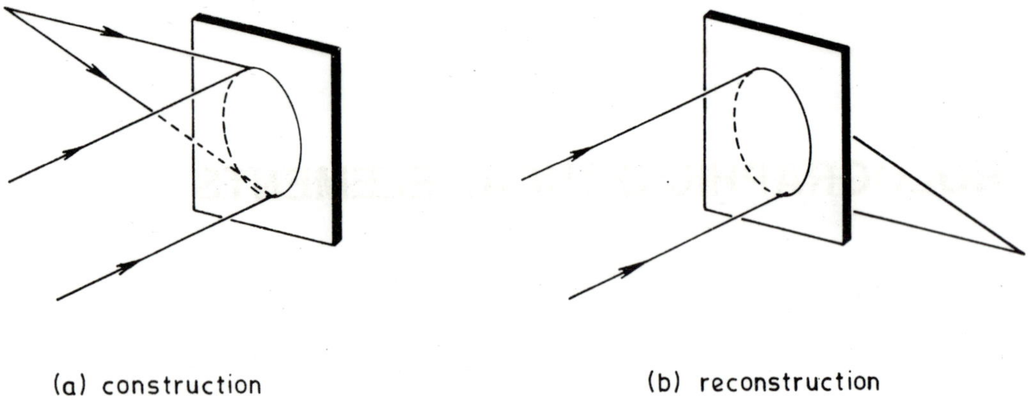

Figure 7.1 Construction and reconstruction of HOE

focal length, aberrations, image orientation, *etc* of a holographic lens vary with wavelength. This dispersive property is undesirable in broadband systems.

(*ii*) A single HOE can serve multiple functions. It can be used as a lens, beam splitter and spectral filter simultaneously (Figure 7.2b).

(*iii*) A single HOE can serve as multiple lenses (Figure 7.2c).

(*iv*) The HOE itself is a thin film of recording material on a substrate. The function of HOE is independent of the shape of the substrate, *i. e.* the optical power does not depend on the substrate. This means that the directions of the entering and exiting rays are not controlled by the substrate normal or its refractive index.

(*v*) In the case of a reflective HOE the substrate need not be of high quality unlike glass lenses which require glass to be highly isotropic, homogeneous and strain-free.

(*vi*) The fabrication and replication of HOEs are relatively easy because no grinding and polishing are required.

(*vii*) The HOE is relatively thin and light-weight.

(*viii*) A holographic element can produce multiple image waves corresponding to different diffraction orders. A lens of primary focal length F has a focal length F/n for the nth diffraction order.

(*ix*) With the availability of real time recyclable recording media any desired system function can be recorded, studied and erased as and when desired.

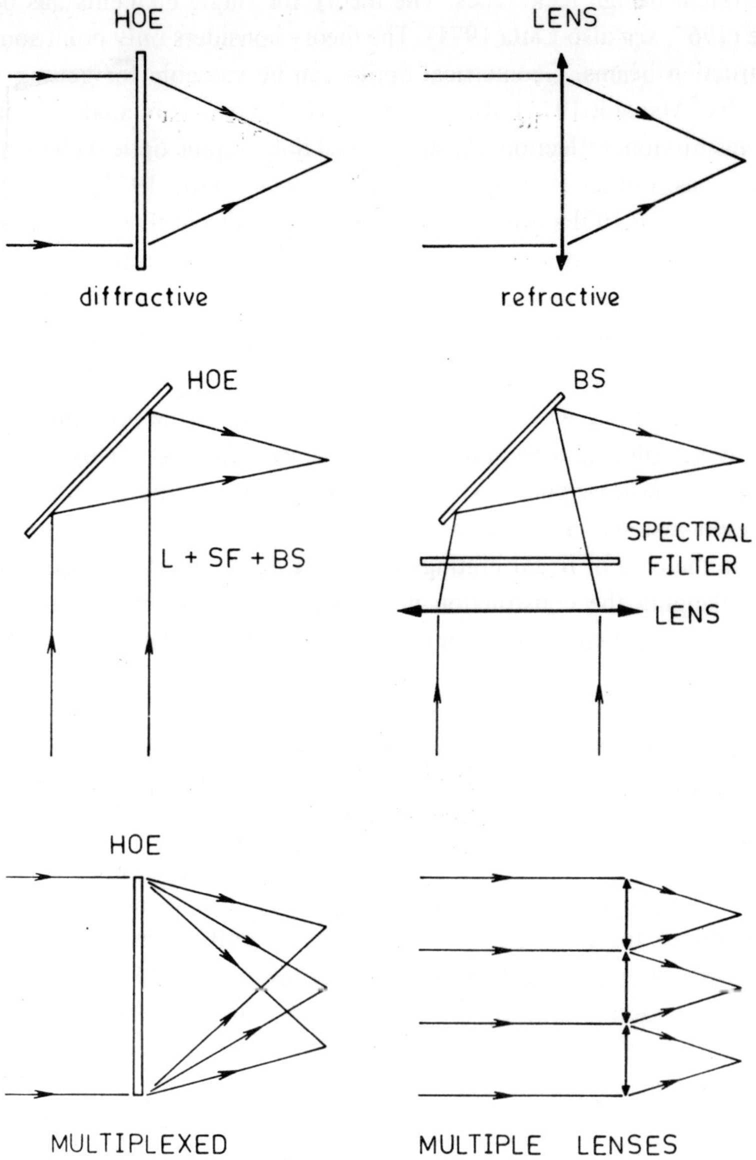

Figure 7.2 Different functions of HOEs: (a) An HOE lens is diffractive whereas glass lens is refractive. (b) A single HOE functioning as lens, spectral filter and beam splitter. (c) A single HOE functioning as multiple lenses.

7.3 Design aspects

Considerable work in optics design has built up a set of starting points for conventional optics. For HOEs no ultimate ready made designs are available so different workers are adopting different design techniques. The theory for single elements has been given by Champagne (1967; *see* also Latta 1971). The theory considers only point source construction/reconstruction beams. Geometrical optics can be valuable for getting first order or paraxial results (Arsnault 1975). Ray tracing of HOE systems in analogy with thin lenses including transmission, reflection, flat and curved holographic optical elements can lead to successful designs (Offner 1966, Close and Mass 1973, Sweatt 1978, Chen 1980). The ray tracing techniques for HOEs are based on the fact that the HOE is a complicated grating for which a generalized vector grating ray tracing equation has been worked out (Close and Mass 1973). A thin lens model (Chen and Wyant 1979) of an HOE can be used with a conventional optical design programme such as ACCOS-5 for designing HOEs.

Lately interest has revived in the wavelength scaling of holographic elements, *i. e.* recording the HOE at one wavelength and extracting the diffraction-limited performance at another wavelength (Latta and Pole 1979, Silvennoinen 1981). This is desired because the recording materials, *viz* photoresists and dichromated gelatine, are sensitive to only the blue/green region whereas applications of HOEs also exist at red and infrared wavelengths. This is easy in the case of linear gratings where the desired performance can be achieved by suitably changing the construction beam angles. For the more complicated optical elements, geometry changes alone would produce an aberrated image. This can, however, be obtained by recording the hologram by aberrated construction beams. Melin (1981) has recently suggested a method which consists in specifying the desired performance at the operating wavelength, selecting one construction beam and utilizing a ray tracing to compute aberrated construction beam required to record the HOE. An HOE capable of focussing this aberrated beam is then designed. A computer-generated hologram can be used to record holograms at visible wavelengths to work at 10.6 μm (Lee 1980). In holographic optical elements the direction of the image ray depends on the recorded fringe structure. Therefore, the grating equation is applicable unlike glass optics where the Snell law is used to determine the direction of the image ray. Two major factors in the design of HOEs are fringe spacing and diffraction efficiency. The fringe spacing determines the image location and the optical power. The diffraction efficiency is governed by the fringe shape and fringe distribution in the recording emulsion.

Let us first consider the recording of a generalized grating. If O exp $(j\,\phi_O)$ and R exp $(j\,\phi_P)$ be the object and reference beams respectively the amplitude transmittance of the recorded hologram will be

$$t = \beta\,[\,O^2 + R^2 + 2\,O\,R\cos(\phi_O - \phi_P)\,]$$

Illuminating the hologram with the readout wave C exp $(j\,\phi_X)$ gives irradiance as

$$I = \beta\left[\,C O^2 E^{j\phi_x} + C R^2 e^{j\phi_x} + C\,O\,R\,e^{j(\phi_X + \phi_O - \phi_P)} + C\,O\,R\,e^{j(\phi_X - \phi_O + \phi_P)}\right]$$

The first two terms in this equation together give the average value of the *dc bias* of the irradiance. The third and fourth terms give the virtual and real images respectively.

Thus the location and shape of the fringes on the hologram are given by the phase difference between the two beams $\phi_O - \phi_P$.

The phase terms and the image point can be calculated by knowing $\phi_i = \phi_C + \phi_O - \phi_R$ from the system geometry. The aberrations can be obtained by comparing the calculated phase with a similar expansion of the phase of a spherical wave.

7.4 Resolution

The resolution of the HOE may be determined by considering the angular spreads of the reference, object and readout wavefronts. The angular resolution of the hologram is limited by these spread functions. The angular spread function of a source such as a pinhole is approximately given by the ratio of the pinhole size to its distance from the hologram. The limiting resolution (LR) is given by

$$LR = \frac{D}{D\,\delta_0 + 2F\,\delta_r}$$

where δ_0 and δ_r are the sizes of the pinholes used for generating the spherical and plane wavefronts respectively, D the aperture of the collimator and F the focal length of the HOE. From this relation it is clear that for high resolution (1) pinholes of the smallest possible

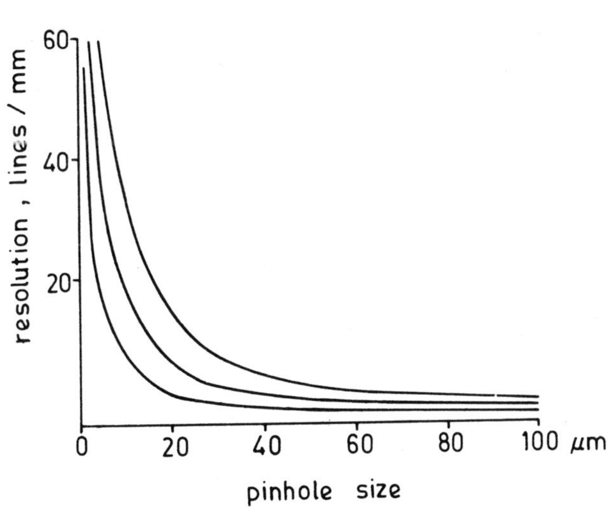

Figure 7.3 Variation of the resolution of HOE fringes with pinhole size for different values of f/-numbers

sizes should be used, (2) the collimator lens and HOE should be as large as possible and (3) the hologram should have the smallest possible *f*/-number.

Figure 7.3 p. 103 shows the variation of resolution with the pinhole size for different values of *f*/-numbers. It is observed that the resolution of the HOE decreases with increase in the *f*/-number and the pinhole size. For a conventional glass lens the resolution is inversely proportional to the *f*/-number. If we use a pinhole of 2μm the resolution will be limited to 310 lines/mm for a lens of *f*/-number 1.

7.5 Aberrations and alignment sensitivity

If the HOEs are not fabricated and used with care they will exhibit many more aberrations than may be present in conventional glass lenses. HOEs suffer from aberrations caused by a change in the wavelength from construction to reconstruction — *chromatic aberrations* — and also by a mismatch, *e. g.* angular or in curvature, in the construction and reconstruction beams — *nonchromatic aberrations*. Both kinds of aberration are quite important even when only small deviations from the recording geometry are present in the reconstruction geometry. The higher order aberrations become important for holograms of large apertures for which alignment tolerances are very low (Méhtā 1974, Méhtā *et al* 1977, Méhtā *et al* 1982). The condition that will eliminate all the aberrations simultaneously is to duplicate one of the construction beams in the reconstruction process. This, however, is very difficult to maintain when HOEs are to be used as components in instruments.

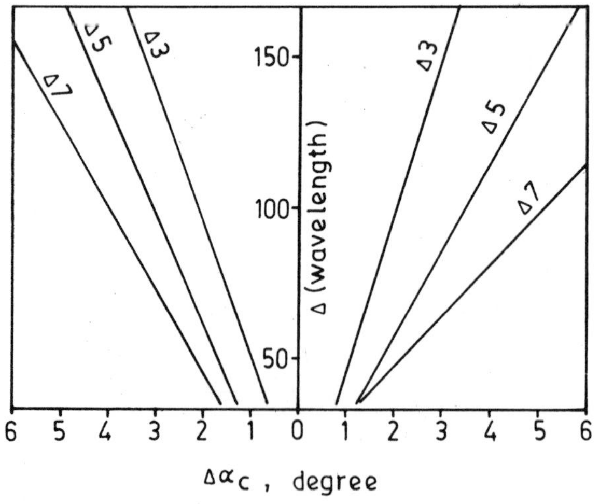

Figure 7.4 Magnitude of the third-, fifth- and seventh-order aberrations (Δ) in an off-axis HOE *vs* the deviation from the reconstruction angle

The third, fifth and seventh order aberrations in an off-axis holographic lens *f*/1 are shown in Figure 7.4 (*see* p. 104). The parameters for the construction of the HOE were: reference and object beam angles with the normal to the HOE 40°, focal length 1 cm and aperture 1 cm. The angle of the reconstruction beam was varied by small amounts. As expected, the magnitudes of the third, fifth and seventh order aberrations are in a decreasing order. The seventh order aberration is quite high even for small deviations in the reconstruction angle.

If we allow permissible aberrations to be less than 5 μm the alignment tolerances can be shown to be very low. For example, the alignment tolerances for comatic and astigmatic aberrations of ≤ 5 μm, reference beam angle of 10°, aperture of 25 mm and focal length of 500 mm are

$$\varepsilon_C \leq 7.2 \times 10^{-5}\ rad \approx 15 \text{ sec } of \text{ arc}$$

$$\varepsilon_A \leq 5.6 \times 10^{-6}\ rad \approx 1.2 \text{ sec } of \text{ arc}$$

The upper limits for tolerances are very low. The HOEs must be mounted in rugged mounts specifically designed to withstand vibrations, shocks and other environmental variations.

7.6 Applications of HOEs

As mentioned earlier, HOEs are able to duplicate most of the functions provided by glass optics only if the optical system operates over narrow spectral bandwidth or requires chromatic dispersion. In general HOEs will not replace glass optics except where their unique characteristics are designed in system applications. Some of the unique functions provided by HOEs are multiple function optics, unusual configurations, narrow spectral response provided by volume reflection holograms, *etc.* Beam combiners, laser scanners, spectral filters, interferometers, gratings, laser collimators and laser schlieren systems are the main areas where typical characteristics of HOEs provide advantages over glass optics.

(a) *Beam combiners*

Beam combiners are important components of various display systems such as head-up display (HUD), helmet-mounted display and night vision goggle systems. Because of the component capability–*viz* imaging, beam splitting and spectral selection–HOEs provide increased flexibility in the selection of image-forming geometry. Conventional display systems are limited by smaller field of view, poor see-through capability and requirement of higher CRT power levels. The HOEs fabricated in dichromated gelatine can overcome all of these difficulties (Chang and Colburn 1977). The visor of the helmet can even function as a combiner plate in helmet-mounted display.

(b) *Laser scanners* (see *Chapter 8 for Holographic Scanners*)

Conventional multifaceted (polygonal) mirror scanners involve high manufacturing cost and it appears difficult to reduce it without innovations. HOEs offer inexpensive replacement of polygonal mirror scanners with added advantages. The optical arrangement of the scanning system can be simpler because of the multiple function capability of HOEs. The speed of the scanning disc can be lowered because the scanning angle depends only on the recorded spatial frequency. Moreover, each "facet" can have a different property (Case and Gerbig 1980).

(c) *HOEs in optical signal processing* (see *Chapter 4 for Optical Signal Processing*)

Holographic lenses can be used as Fourier transforming lenses and complex filters in optical signal processing experiments (Méhtā *et al* 1977). A compact optical data processor can be built using holographic reflective lenses, requiring only one-half of the length of the optical bench as that required for using glass lenses (Figure 7.5). The processor can be used for various optical signal processing experiments, *viz* halftone/raster removal, frequency doubling, separation of two signals, directional filtering, contrast inversion, matched filtering, *etc*.

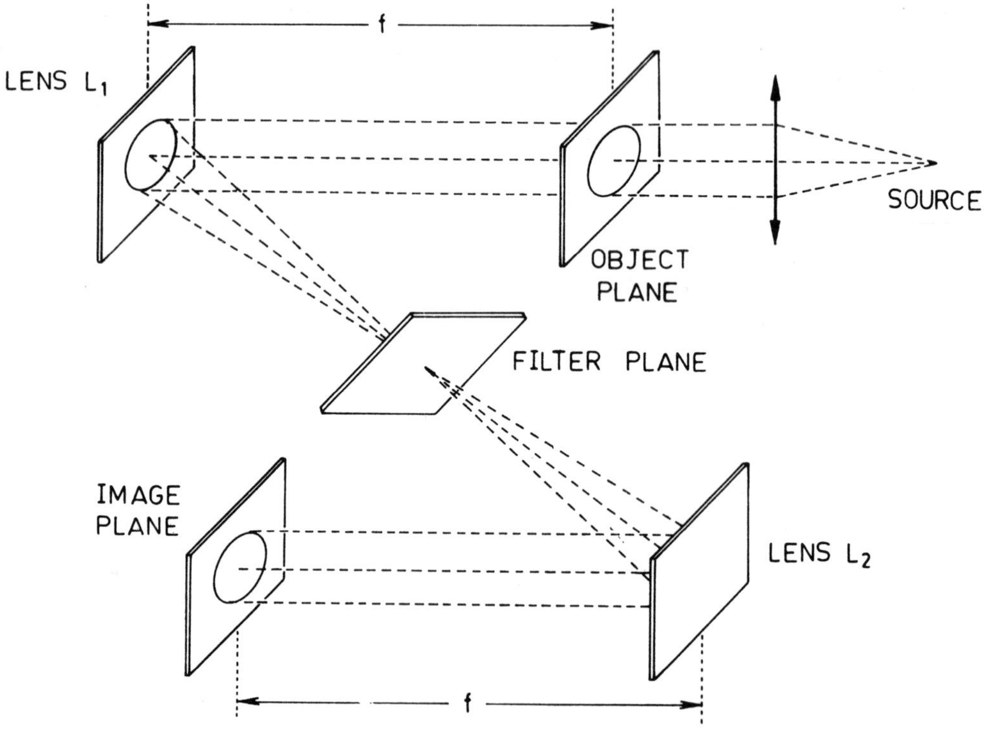

Figure 7.5 Optical data processor employing holographic reflective lenses

Matched filtering is one of the most powerful operations that can be performed by optical data processors using holographic matched components. HOEs are bound to play an important role in optical pattern recognition systems of compact size to be used in airborne military applications (Shen *et al* 1980). Holographic gratings and phase plates are used for image multiplication (Méhtā 1975, Méhtā *et al* 1979) and coherent noise elimination (Katti and Méhtā 1976, Méhtā *et al* 1979).

(d) *Fibre optics and optical waveguide components*

Holographic gratings and lenses find applications as couplers in integrated optics and fibre optics. Holographic multiplexers are ideal for wavelength division multiplexing in fibre optical systems. This is because of the fact that a single HOE performs all of the functions of collection, wavelength separation and focussing. The properties of HOE couplers are as follows:

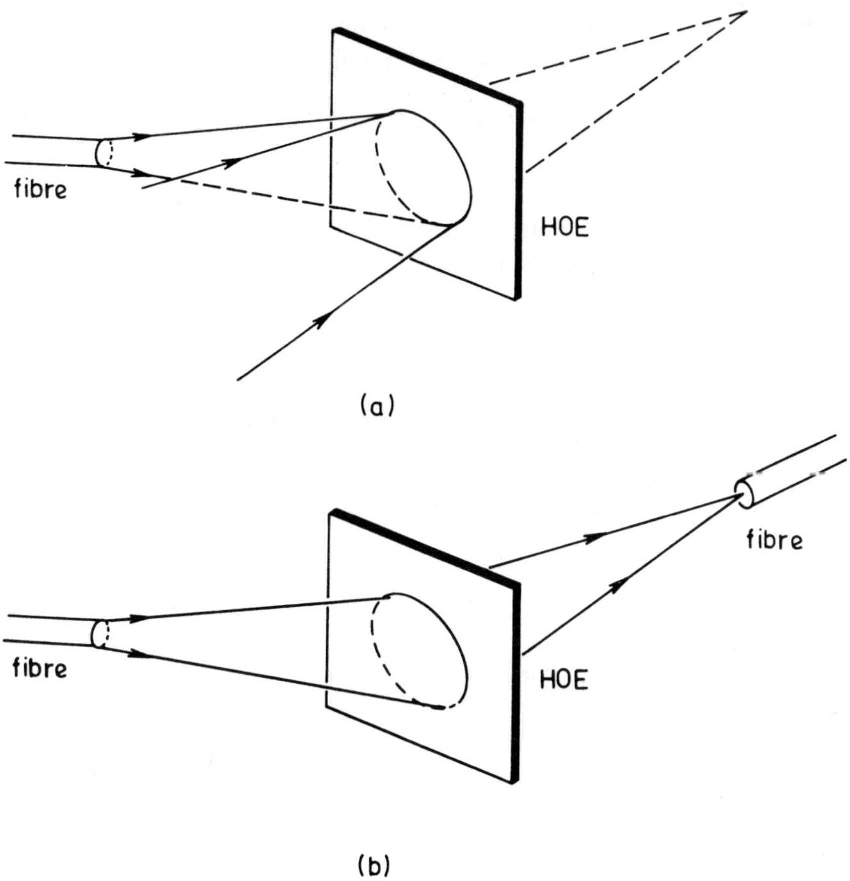

Figure 7.6 Holographic fibre coupler: (a) Recording, (b) Coupling

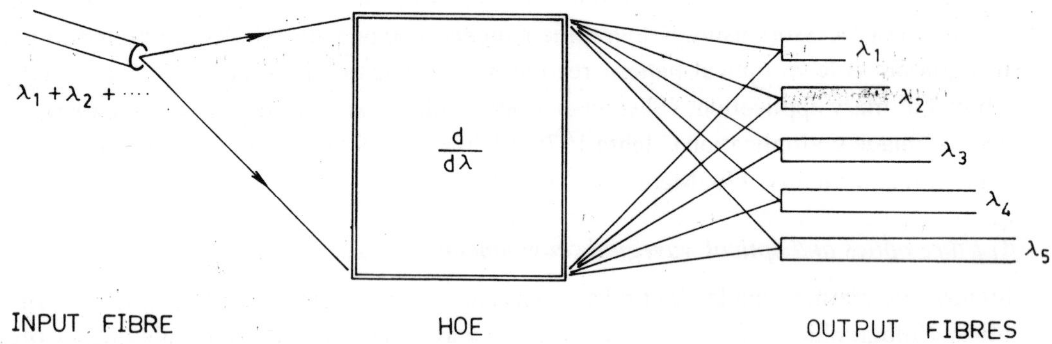

Figure 7.7 Wavelength division holographic demultiplexer

(*i*) Easy fabrication and replication of any *f*/-number and focal length

(*ii*) Mode-conversion holocouplers can be made for the conversion of the spherical output of a fibre to a form suitable for a thin-film integrated circuit.

(*iii*) A fibre can be coupled to multiple fibres through a multiplexed holocoupler.

(*iv*) Two fibres at any angle can be coupled.

Figure 7.6 shows the principle of the holographic coupler. The use of HOE as a wavelength multiplexer is shown schematically in Figure 7.7.

7.7 HOE Fabrication materials

The full exploitation of the potentialities of HOEs depends on the availability of high quality hologram recording materials to achieve high optical efficiency and image quality. The required characteristics are as follows:

(*i*) Peak diffraction efficiency above 90% at the working λ

(*ii*) High spectral sensitivity at the recording wavelength

(*iii*) High contrast-recording of hologram fringes

(*iv*) High resolution (at least 5000 lines mm^{-1})

(*v*) Resistance to environmental changes

(*vi*) Negligible emulsion shrinkage

(*vii*) Low scattering and absorption losses

(*viii*) Adequate thickness

7.8 Conclusion

Holographic optical elements today have matured to a stage where it is possible to have a number of HOE-based instruments. Holographic night vision goggles and holographic head-up display systems have already been introduced in the defence services in many countries. It is expected that all futuristic motor cars equipped with HUDs will become a reality in the coming decade. Holographic scanners are commercially available. Specialized HOEs are being thought of which would be useful as routing elements and interconnects in futuristic all-optical computers and for increasing the speed of the present day digital computers. The HOEs will be dynamic and programmable and be able to send a signal from a source to a number of points thus making communication easy between wafers or chip-to-chip or within a chip.

The present day HOEs are being recorded mainly in dichromated gelatine emulsion. However, this emulsion is not available in precoated form and one has to prepare one's own plates. The final HOE is highly dependent on the environmental conditions existing at the recording stage. The finished HOE is affected by ambient humidity and temperature. Environmental effects can be eliminated by proper sealing of the hologram. Efforts are being made to replace dichromated gelatine. Photopolymers are considered to be a good substitute for dichromated gelatine. HOE recording in polymers is not affected by environmental conditions, neither at the recording time nor while in use. Though further research is called for, the present stage of development is adequate to produce high quality HOEs.

References

ARSNAULT H H, 1975 J Opt Soc Am **65** 903

CASE S K and GERBIG V, 1980 Opt Engg **19** 711

CHAMPAGNE E B, 1967 J Opt Soc Am **57** 51

CHANG B J and COLBURN W S, 1977 J Opt Soc Am **67** 1373

CHEN C W, 1980 Opt Engg **19** 649

CHEN C W and WYANT J C, 1979 J Opt Soc Am **69** 1402

CLOSE D H and MASS G E, 1973 J Opt Soc Am **63** 1524

KATTI P K and MÉHTĀ P C, 1976 Appl Opt **14** 530

LATTA J N, 1971 Appl Opt **10** 599, 609, 2698

LATTA M R and POLE R V, 1979 Appl Opt **18** 2418

LEE W H, Opt Comm **34** 29

MALIN H, 1981 Opt Engg **20** 756

MÉHTĀ P C, 1974 Opt Acta **21** 1005, 1975 Opt Comm **13** 40

MÉHTĀ P C, BHAN C and HRADAYANĀTH R, 1979 J Opt (Paris) **10** 133

MÉHTĀ P C, Mohan D and HRADAYANĀTH R, Opt Comm **288** 171

MÉHTĀ P C, Râo K S S and HRADAYANĀTH R, 1982 Appl Opt **21** 4553

MÉHTĀ P C, SWAMI S and RĀMPĀL V V, 1977 Appl Opt **16** 445

OFFNER A, 1966 J Opt Soc Am **56** 1509

SHEN M, Casasent D, LUU T K and FENG B, 1980 Opt Comm **34** 311, 316

SILVENNOINEN R, 1981 Opt Comm **39** 216

SWEATT W C, 1978 Appl Opt **17** 1220

Chapter 8

HOLOGRAPHIC SCANNERS

P C MÉHTĀ

8.1 Introduction

Holography has not remained merely a technique for recording 3D images but has grown into a new subject having potential applications in diverse fields. Ever since the demonstration of 3D images the interest had been aroused in creating highly realistic and accurate 3D images but that could not be possible until the mid-1960s because of the lack of a suitable source of light. The major upsurge of interest in holography was triggered off by the development of the HeNe laser and the *off-axis* technique of Leith and Upatnieks (*see* Chapter 5) that paved the way for recording 3D scenes realistically. Fourier transform holography (*loc cit*), holographic interferometry (*see* Chapter 6), rainbow holography (*see* Chapter 5) and holographic optical elements (*see* Chapter 7) have opened up new areas of application. Besides 3D displays holography has now been accepted as a viable tool for non-destructive testing of materials, stress-strain analysis, diagnostics of automotive industry, optical testing, encoding of information, particle sizing and optical signal processing.

Laser beam scanning (Figure 8.1-6*) is one area where the full potential for complex wavefront recording and reconstruction has been exploited commercially. *Holographic Scanners* (*HSs*) have already proven their usefulness for reading the universal product (bar) code from merchandise in supermarkets (Figure 8.6). They have had an impact on laser printing and copying as well as on laser materials processing also.

Before discussing the details it will be useful to see the reasons for considering holograms for light deflection applications when a number of methods already exist such as rotating mirrors and acousto-optical scanners. Following are the points in favour of holograms:

* For convenience all figures have been inserted at the end.

(*i*) Focussing in addition to deflection can be combined in a single hologram.

(*ii*) Holograms can be used in two modes: as deflectors generating moving spots of light and as scanners collecting some portion of the light that is scattered by some object scene. An important feature of the retroreflection mode is that the location of the return image is invariant with the position of the moving spot in the scan field.

(*iii*) Resolution enhancement beyond the limits set by the numerical aperture of the scanning beam becomes possible by detector plane filtering.

(*iv*) The hologram serves as an ambient light filter, directing only light within the proper spectral bandwidth to the detector plane.

(*v*) The size and the weight of the hologram are negligible, making the drive system easier.

(*vi*) Holograms can be mass-replicated.

Holographic scanners are mechanically moving deflectors, like rotating mirrors, but have smooth surfaces and are, therefore, suitable for faster movement with lower drive power and better speed control because device distortion due to centrifugal forces is negligible, and so is turbulence in the surrounding atmosphere. HSs do not suffer from beam displacement and nonlinear deflection exhibited by some mirror systems. Further, HSs are highly wobble-insensitive in transmission. Their most important advantage over mirrors is their scan flexibility.

HSs are 2D devices with 3D scan possibility and thus are more flexible than acousto-optical scanners.

The factors which are against holograms for their use in laser beam scanners are their lack of spectral broad-bandedness and a lack of high efficiency holographic materials sensitive to red and infrared wavelengths.

8.2 Hologram types

Holographic scanners are based on the diffraction of light by varying holographic fringe patterns. This is usually done by translating or rotating holograms illuminated by spatially coherent and monochromatic laser light. Only a few types of hologram are useful for HSs since high diffraction efficiency and ease of production are the main criteria. Ruling out the light-absorbing holograms of efficiency smaller than 10% leaves us with *phase-only* structures due to refractive index variation or surface profile. Some other considerations may be listed as follows:

(*i*) Scalar diffraction theory tells us that the highest efficiency for thin, sinusoidally modulated, phase-only holograms is 34% and for binary holograms 40%. Such holograms are fairly easy to produce and can be used for HSs, especially if both first orders are used.

(*ii*) Close to 100% efficiency can be achieved by diffraction of polarized light from surface relief gratings of high spatial frequencies and proper profile depth. Such gratings can be replicated at very low cost.

(*iii*) Blazed gratings give 100% efficiency, but they are very difficult and expensive to manufacture.

(*iv*) Volume phase holograms are another important class. Dichromated gelatine can give 100 per cent efficiency.

8.3 Optical considerations

The choice between using a transmission hologram and using a reflection one is dependent on the required scan line uniformity and repeatability. In a transmission hologram, image quality and scan line repeatability are strongly influenced by the quality and uniformity of the substrate. Parallelism of the two surfaces comprising the substrate and possible protecting covering is a stringent demand. Any wedge angle α will introduce a deviation $\Delta\theta$ into an incident beam given by

$$\Delta\theta = (n - 1)\,\alpha$$

where n is the refractive index of the substance. For example, a total wedge angle of $1'$ causes a total deviation of 73 μm in one full disk revolution in scan line location for a 500 mm working distance. To achieve a deviation of less than ± 2.5 μm the wedge angle requirement becomes 2 arcsec.

Both transmission and reflection holograms are subject to errors introduced by a lack of orthogonality between the rotation axis and the hologram plane. This can be a mixed error due to incorrect mounting and bearing wobble. This introduces cross-scan displacement. The deviation $\Delta\theta$ in the diffraction direction for small tilt angle α of the diffracting surface with respect to the illumination direction is given by

$$\Delta\theta = \left[\frac{\cos\theta}{\cos\phi} \mp 1\right]\Delta\alpha \pm \left[\frac{\sin\phi - \sin\theta}{2\cos\theta}\right](\Delta\alpha)^2$$

where the upper sign is for a transmission grating and the lower for a reflection one. Thus a reflection hologram behaves like a mirror while beam deviation in the transmission hologram is a smaller, second order effect.

These considerations present two primary options. The first is to use a transmission hologram so as to exploit its wobblefree characteristic and to minimize residual cross-scan effects by using good optical quality substrates. The other is to use a reflection hologram that can be fabricated on a variety of opaque and transparent substrates.

Any angular mismatch between the original recording direction and the subsequent illumination direction will result in a loss of diffraction efficiency, a mismatch of the diffracted beam and/or the introduction of wavefront aberrations. It is possible, however, to maintain the required angular alignment by selecting a configuration for which the optimum location of the reference and the playback sources is invariant with hologram motion. An example of such a motion is rotation. If the centre of curvature of the reference source is positioned on the axis of rotation a similarly directed illumination beam will have the proper incidence for every holographic subarea during its rotation.

8.4 Parameter definitions

Design parameters include spot size, scan length, scan straightness, scan linearity, field curvature, scan speed and latency. The resolution is decided by the spot size and scan length. Scan straightness is very application-dependent; for example, it is highly important for a laser printer with a cylindrical photodetector and relatively unimportant in a label scanner.

Scan linearity is defined as a one-to-one correspondence between a unit increment of time during the scan and the incremental displacement of the beam in the scan plane. Lack of linearity will cause nonuniform recording when using a fixed frequency data clock and hence will require real time clock correction technique.

Scan latency refers to that fraction of time during which the scanning beam is unusable (*see* Figure 8.4, for example). This is caused by the finite size of the beam illuminating the hologram and its overlap on to adjacent holograms at their common boundary. This parameter, often referred to as *dead time*, is similar to the flyback time in cathode ray tube technology and greatly influences data rate and throughput. Excessive latency makes inefficient use of the scanner and may require the use of high burst data rates to accomplish the task.

Field curvature has an impact on image spot quality and scan line linearity.

Diminished spot quality, as characterized by spot size nonuniformity across the scan plane, results from deviation of the focal waist location from a true plane. (*See* Chapter 12 for *focal waist*.)

Scan speed is one of the more independent parameters, having a reasonable degree of freedom. Although it is determined by both the number of facets and rotational speed, the latter is constrained only by mechanics and strength of materials.

Both random and schematic errors in the scan line spacing and pixel placement within the lines will degrade image quality. Scan line spacing errors are caused by either photoreceptor velocity errors or cross-scan positional errors — termed *tracking errors* — associated with the deflector system. In polygonal mirror systems the tracking error is due primarily to facet-to-facet tilt alignment error and deflector wobble.

Pixel placement error within scan lines is termed *jitter* and is most noticeable as waviness of lines running perpendicular to the scan direction. In rotational scanners such as polygon and *hologon* the synchronization errors are primarily caused by scan spot velocity inaccuracies resulting from errors in the rotational speed, facet location and surface figure.

8.5 Scan tasks

We call scan pattern that patch of light which is directed along the scan path. The path can be in one, two or three dimensions of space, continuous or not, or even consisting of discrete points. It can be scanned fast or slow, forth or back, with linear speed or not, once only or repeatedly cyclic or oscillatory. The pattern can be only one pixel large (the diffraction limit) or spread out (in 1D or 2D). It can be of some symmetry and orientation or not, bright or dim, monochromatic or not, of constant shape and orientation and brightness and colour or not.

8.6 Scanner types

Depending on the application and accuracy required, single or a number of holograms can be used in a scanner. Following are the possibilities:

(a) *Single hologram scanners*

- (*i*) Rotating gratings (Figure 8.1,8.2)
- (*ii*) Rotating holographic lenses
- (*iii*) Rotating general holograms
- (*iv*) Translating gratings
- (*v*) Translating holographic lenses (Fig 8.3)
- (*vi*) Translating general holograms (Fig 8.5)

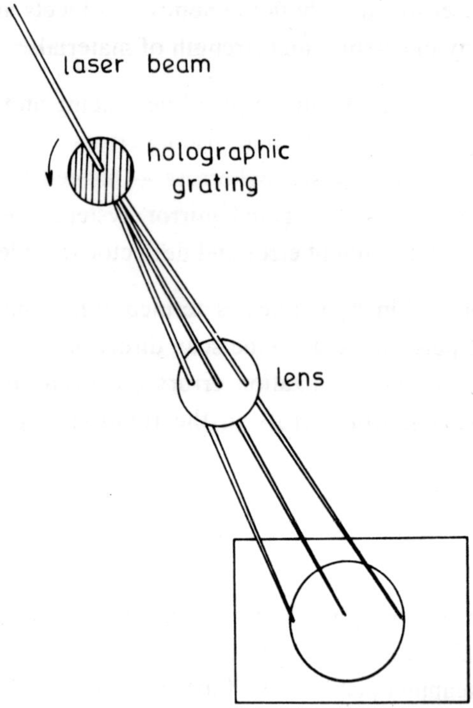

Figure 8.1 Scanning by a rotating holographic grating: laser beam focussed by a convex lens

(b) *Tandem hologram scanners*

(*i*) Counter-rotating gratings

(*ii*) Counter-translating Fresnel zone plates

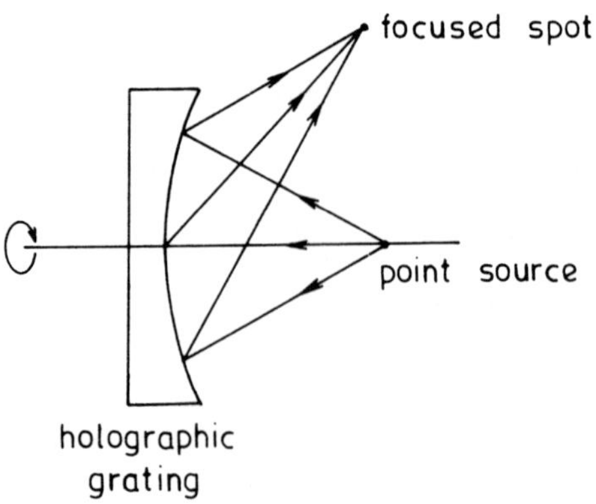

Figure 8.2 Scanning by a holographic concave grating: point source focussed into a spot

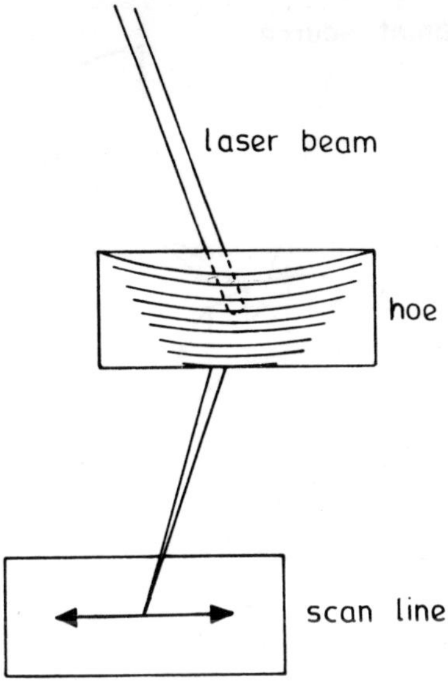

Figure 8.3 Scanning by translating HOE

(*iii*) Faceted tandem hologram scanner

(*iv*) Anamorphic tandem hologram scanner

8.7 Holographic media

The selection of deflector type and configuration must take into account the compatibility of available holographic materials and process. In addition, the application itself may indicate the holographic material and configuration. This is particularly true of those applications where the hologram is used bidirectionally, as in UPC scanners, and hence which require higher diffraction efficiency.

The most promising materials for holographic deflectors are dichromated gelatine, surface relief in photoresist, and bleached holographic emulsion. The ultimate choice will depend upon the efficiency required, cost and manufacturability.

8.8 Manufacturing

The choice of manufacturing techniques is limited to either fabrication of every scanner as an original master or the creation of one master recording and subsequent volume production by the generation of multiple copies. The primary factors in selecting a process are the operating wavelength, reconstruction efficiency, need for design flexibility, manufacturing throughput and product cost.

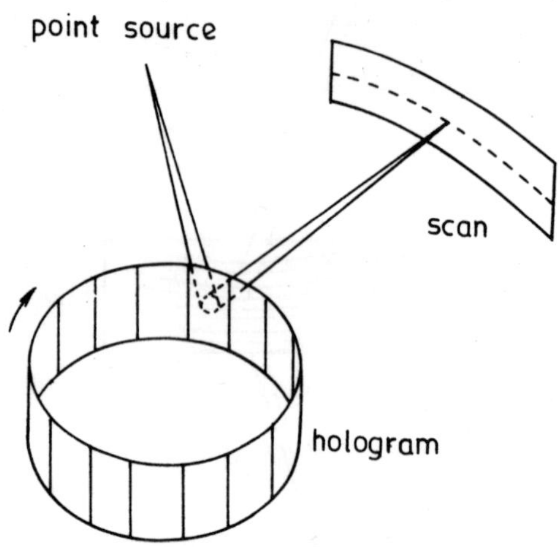

Figure 8.4 Scanning by rotating faceted hologram

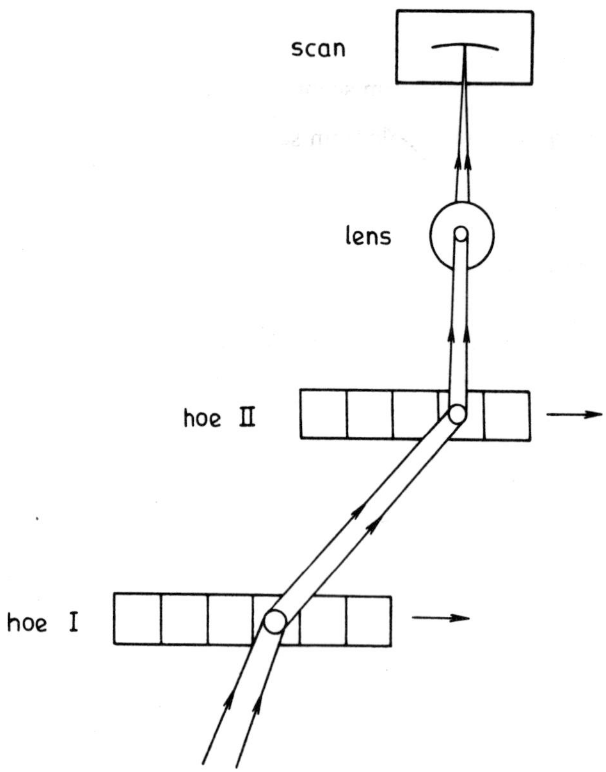

Figure 8.5 Scanning by two translating HOEs

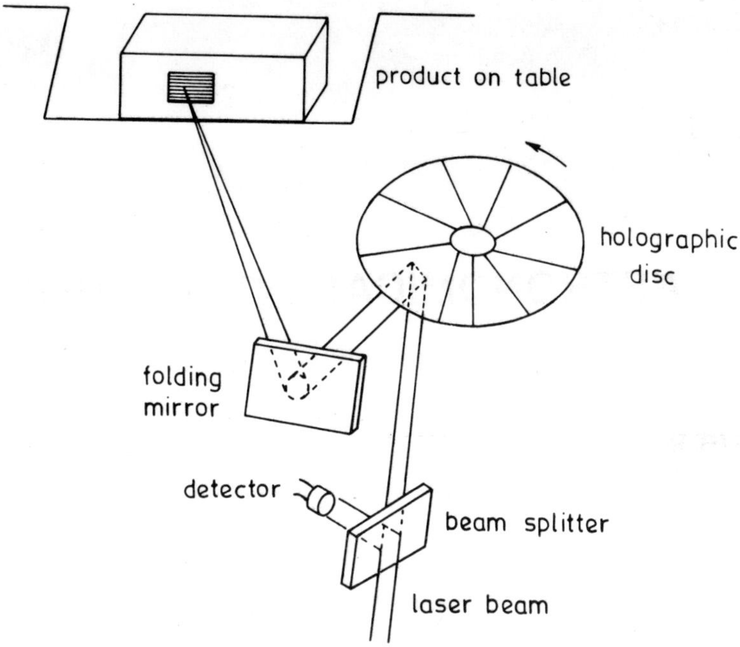

Figure 8.6 Schematic digram of UPC scanner

8.8.1 The possibilities are

Production by direct recording of transmission or reflection holograms

Production by replication by optical copying or mechanical replication

8.8.2 Practical Systems

The concept of using holographic scanning has been around for more than two decades and during this time many different applications have been suggested, but only a few have been demonstrated and an even smaller number has made it to the market place. There are only five announced commercial products: point-of-sale scanners by IBM, Nippon Electric Company and Fujitsu, and electrophotographic printers by Benson and General Optronics.

Suggested Reading

Ih C, 1978 Appl Opt **16** 2137, 1978 **17** 748

Ikeda H *et al*, 1979 Appl Opt **18** 2166

Kramer C, 1981 Laser Focus **17**(6) 70

Marshall G F, 1985 *Laser Beam Scanning*, M Dekker & Co

Chapter 9

RADIO ASTRONOMICAL INTERFEROMETERS

N C MĀTHUR

9.1 Introduction

Radio astronomy is the study of astronomical objects at radio wavelengths. The instrument used is called a *radio telescope* and consists of a high gain, narrow beam antenna associated with a low noise receiving system. Astronomical objects such as the sun, planets, radio galaxies, pulsars, quasars, *etc* are extremely weak sources of radiation. The flux density is measured in terms of the unit *jansky* named after Karl Jansky who discovered radio noise from the milky way galaxy in 1932. One jansky is equal to 10^{-26} W m^{-2} Hz^{-1}. Due to long wavelengths used in radio astronomy compared with optical wavelengths the size of the antenna has to be very large to achieve a reasonable resolution. At 1 cm wavelength, for example, an antenna of diameter 80 m will have the same resolving power as the human eye while an antenna having resolution comparable to that of the *Venu Bappu* optical telescope at Kovalur (diameter 254 cm) would have to be over 40 Km in diameter! Clearly it is impossible to build single antennas large enough to achieve resolutions comparable to optical telescopes.

Radio interferometers offer a unique method of achieving high resolution in radio astronomy. Called *Synthetic Aperture Radio Telescopes*, they can achieve resolutions better than 1 arcsec. The *Very Long Baseline Interferometry* technique has been used successfully to achieve resolution up to 0.001 arcsec. Moreover, interferometers can be used to obtain detailed maps of the brightness distribution of extended sources. The theory of interferometry slightly and aperture synthesis is discussed below.

9.2 Theory of interferometry

9.2.1 Interference and fringes

Consider a source S the radiation of which is observed at a point Q through two apertures in an infinite plane. Let $U_1(t)$ and $U_2(t)$ be the fields produced at the apertures P_1 and P_2 respectively (Figure 9.1). The total field at Q then is

$$U(Q,t) = K_1 U_1\left(t - \frac{d_1}{c}\right) + K_2 U_2\left(t - \frac{d_2}{c}\right) \quad 9.1$$

K_1, K_2 depend on aperture size and source location with respect to P_1, P_2. In general, $K_i \propto (1/d_i)$. The irradiance $I(Q)$ at Q is defined, following Born and Wolf (1965), by

$$I(Q) = \langle U(t)\, U^*(t) \rangle$$

$$= \lim_{T\to\infty} \left[\frac{1}{2T} \int_{-T}^{T} U(t)\; U^*(t)\, dt \right] \quad 9.2$$

The mutual coherence function, defined as

$$\Gamma_{12}(\tau) = \langle U_1(t+\tau)\; U_2^*(t) \rangle \quad 9.3$$

is the crosscorrelation of U_1 and U_2. The autocorrelation functions of U_1 and U_2 are defined as

$$\Gamma_{11}(\tau) = \langle U_1(t+\tau)\, U_1^*(t) \rangle \quad 9.4$$

$$\Gamma_{22}(\tau) = \langle U_2(t+\tau)\, U_2^*(t) \rangle$$

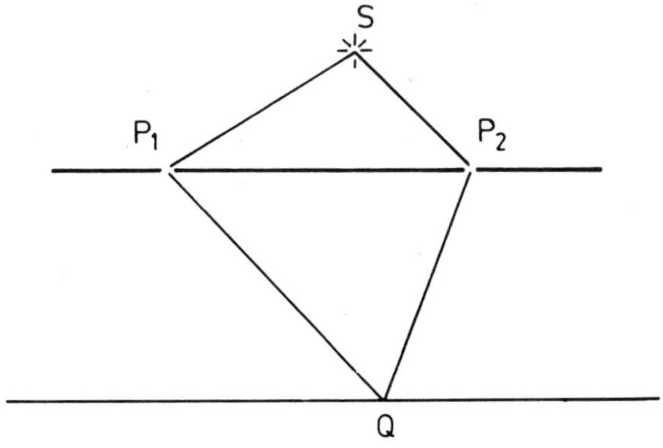

Figure 9.1 Interference Geometry

Clearly, for $\tau = 0$, we have

$$\Gamma_{11}(0) = I_1(Q), \; \Gamma_{22}(0) = I_2(Q)$$

The complex degree of coherence is defined as

$$\gamma_{12}(\tau) = \frac{\Gamma_{12}(\tau)}{\sqrt{\Gamma_{11}(0)\,\Gamma_{22}(0)}} = \frac{\Gamma_{12}(\tau)}{\sqrt{I_1 I_2}} \qquad 9.5$$

It may easily be shown that the total irradiance at Q is given by

$$I(Q) = I^{(1)}(Q) + I^{(2)}(Q) + 2\sqrt{I^{(1)}(Q)\,I^{(2)}(Q)}\,\mathrm{Re}[\gamma_{12}(\tau)] \qquad 9.6$$

where $I^{(i)}$ (Q) is the irradiance at Q due to P_i alone. The irradiance at Q depends upon the location of the source and the point of observation relative to P_1 and P_2 and upon the nature of the source. The source is said to be completely coherent, partially coherent or completely incoherent accordingly as

$$|\gamma_{12}(\tau)| = 1, < 1, 0 \qquad 9.7$$

For a monochromatic source of frequency ω

$$\mathrm{Re}[\gamma_{12}(\tau)] = \cos(\omega t - \phi_{12}), \; |\gamma_{12}(\tau)| = 1 \qquad 9.8$$

where $\tau \equiv (d_1 - d_2)/c$ and ϕ_{12} is the phase of the field in aperture P_1 relative to that in P_2. For the monochromatic case the irradiance is given by

$$I(Q) = I^{(1)}(Q) + I^{(2)} + 2\sqrt{I^{(1)}(Q)\,I^{(2)}(Q)}\,\cos(\omega\tau - \phi_{12}) \qquad 9.9$$

Thus, as the source moves (thereby changing ϕ_{12}) or the point of observation Q moves in the image plane (thereby changing τ), the observed irradiance varies sinusoidally between maximum and minimum values given respectively by

$$I_{\max} = I^{(1)}(Q) + I^{(2)}(Q) + 2\sqrt{I^{(1)}(Q)\,I^{(2)}(Q)} = (U_1 + U_2)^2 \qquad 9.10$$

$$I_{\min} = I^{(1)}(Q) + I^{(2)}(Q) - 2\sqrt{I^{(1)}(Q)\,I^{(2)}(Q)} = (U_1 - U_2)^2$$

In astronomical situations $|U_1| = |U_2| = |U_0|$ and the irradiance varies sinusoidally between zero and $(2U_0)^2 = 4I_0$ for the monochromatic case. In the *quasi-monochromatic* case, where the source radiation has a small but finite bandwidth $\Delta\omega$ such that $[(\Delta\omega)/\omega] << 1$ we have $\mathrm{Re}(\gamma_{12}) < 1$ and the total irradiance varies between the limits $2(1 + |\gamma_{12}|)$ I and $2(1 - |\gamma_{12}|)$ I. This is the partial coherence case. For complete incoherence $\gamma_{12} = 0$ and the total irradiance is constant. These situations are shown in Figure 9.2 (p. 123). The successive maxima and minima are called *fringes* and *fringe visibility* is defined as

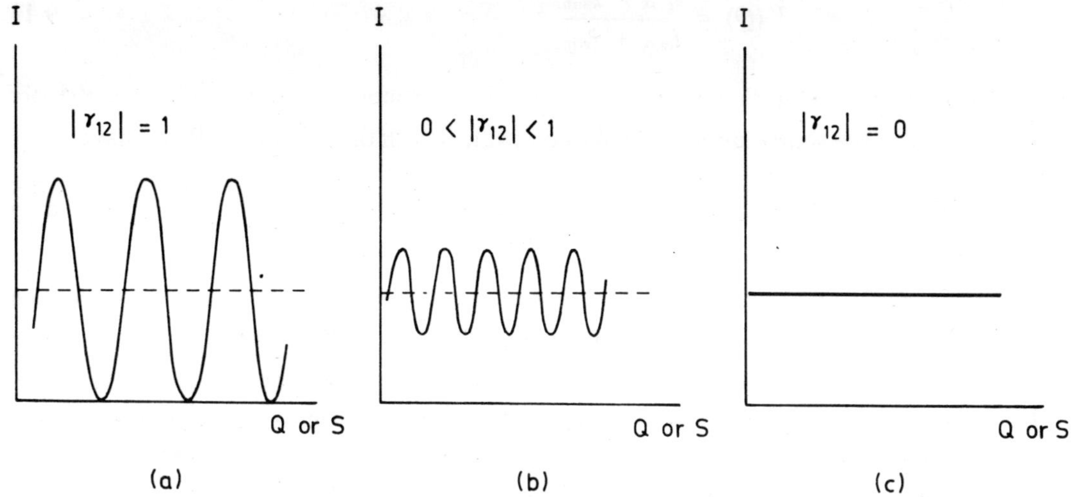

Figure 9.2 Fringe response for different degrees of coherence, (a) Completely coherent, (b) Partially coherent, and (c) Completely incoherent

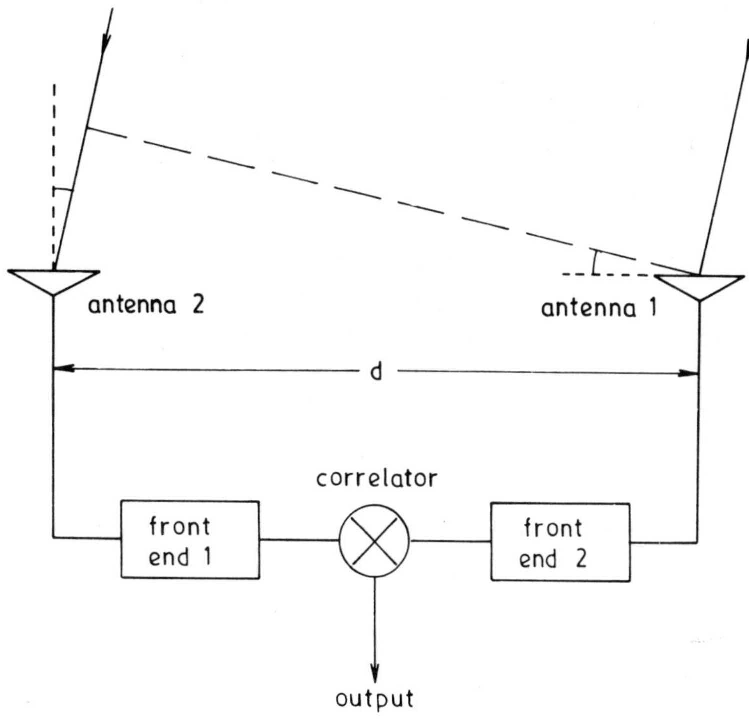

Figure 9.3 Basic Correlator Interferometer

$$V(Q) = \frac{I_{max} - I_{min}}{I_{max} + I_{min}} \qquad 9.11$$

It can easily be shown that V(Q) = $|\gamma_{12}|(\tau)$. For astronomical sources P_1S and P_2S are parallel and if the source direction makes an angle θ with the *baseline* P_1P_2 we have

$$\phi_{12} = \omega \frac{d\cos\theta}{c} = 2\pi \frac{d\cos\theta}{\lambda} \qquad 9.12$$

When *d* is large compared to λ the successive maxima of the fringe pattern occur at an angular spacing of $\lambda/d \sin\theta$. This is called *fringe spacing* and is inversely proportional to the projection of the baseline normal on the source direction measured in wavelengths. Fringes are produced due to successive constructive and destructive interference between the rays received from the two apertures as the source moves in the sky. It is clear, therefore, that for an extended source (as against a point source) radiation from different parts of the source does not add or cancel at the same time. This will result in a reduction in the fringe visibility. If the source size exceeds the fringe spacing the fringes will disappear completely. The situations are similar to the partially coherent and completely incoherent cases respectively. Thus the existence of the fringes puts an upper limit on the size of the source. If fringes do occur the source is said to be *resolved.*

9.2.2 The correlator interferometer

In the correlator interferometer, extensively used in radio astronomy these days, a source is observed by two antennas simultaneously and the outputs of the two receivers are correlated. The theory of the correlator interferometer is given below and it is shown that in special cases (which obtain in most radio astronomical situations) the correlator output measures fringe visibility which is one component of the Fourier transform of the source brightness distribution. This forms the basis for aperture synthesis.

9.2.2.1 Generalized response

The signals received by the two antennas are processed through the front ends of the receivers (consisting of low noise amplifiers and mixers) and the intermediate frequency (IF) signals are crosscorrelated as shown in Figure 9.3 (p. 123). In the idealized case of a monochromatic point source the two signals being correlated are identical sine waves except that the one coming from antenna 1 is delayed by an interval $\tau = (d\sin\xi)/c$ with respect to that from antenna 2. Thus the correlator output varies as cos (2 ω τ). For short intervals of time, τ varies linearly with time and the output shows fringes with a fringe spacing $\lambda/(d\cos\xi)$ radian. In actual practice we deal with extended sources with a wide frequency spectrum and the antennas and front ends contaminate the signals before correlation. A mathematical model incorporating these features is as follows (Swenson and Māthur 1968).

Let $E(\xi,t)$ represent the field from a source in the direction ξ at time t. For simplicity we consider the one-dimensional case. (In general, the source direction will be represented by the direction cosines.) Although radio astronomical signals are not timebound we shall assume that the temporal frequency spectrum as well as the spatial frequency spectrum of $E(\xi,t)$ exists (*see* Born and Wolf *loc cit* p 497 for the truncated function approach to the existence of temporal frequency Fourier transforms). We shall represent temporal frequency domain quantities with a circumflex and spatial frequency domain quantities in small case letters. Thus, corresponding to the signal $E(\xi, t)$, there are two other quantities defined through Fourier transforms as follows:

$$\hat{E}(\xi, \nu) = \int_{-\infty}^{\infty} E(\xi, t)\, e^{-j2\pi t}\, dt$$

$$e(u, t) = \int_{-\infty}^{\infty} E(\xi, t)\, e^{j2\pi u\xi}\, d\xi \qquad 9.13$$

being the temporal and spatial frequency spectra of the source brightness distribution respectively. The inverse Fourier transforms are given by

$$E(\xi, t) = \int_{-\infty}^{\infty} \hat{E}(\xi, \nu)\, e^{j2\pi t\nu}\, d\nu$$

$$E(\xi, t) = \int_{-\infty}^{\infty} e(u, t)\, e^{-j2\pi u\xi}\, d\xi \qquad 9.14$$

The mutual coherence function of the signals from two different directions is defined as

$$\Gamma(\xi_1, \xi_2, \tau) \equiv \langle E(\xi_1, t)\, E^*(\xi_2, t+\tau) \rangle \qquad 9.15$$

Similarly, the *mutual coherence function of the spatial frequency spectrum* is defined as*

$$\gamma(u_1, u_2, \tau) \equiv \langle e(u_1, t)\, e^*(u_2, t+\tau) \rangle \qquad 9.16$$

For incoherent sources (hence applicable to radio astronomy) radiations from two different parts of an extended source are uncorrelated and we have

$$\Gamma(\xi_1, \xi_2, \tau) = \Gamma(\xi_1, \tau)\, \delta(\xi_1 - \xi_2) \qquad 9.17$$

* These Γ and γ are an FT pair, *not* the earlier Γ and γ. The context should make it clear.

Clearly Γ (*u*, τ) and γ (ξ, τ) form an FT pair:

$$\Gamma(\xi, \tau) = \int_{-\infty}^{\infty} \gamma(u, \tau)\, e^{-j2\pi\xi u}\, du$$

$$\gamma(u, \tau) = \int_{-\infty}^{\infty} \Gamma(\xi, \tau)\, e^{j2\pi u\xi}\, d\xi \qquad 9.18$$

With these definitions we proceed to find the response of the correlator interferometer. The contamination of the signal by the antennas and front ends mentioned earlier can be mathematically modelled as follows. An antenna does not receive radiation equally well from all directions and its relative effectiveness as a function of direction is given by its radiation pattern $\hat{G}(\xi, \nu)$ — where the dependence on frequency θ is only marginal over a small bandwidth. The response of an antenna to an extended source distribution is given by the convolution of the source distribution and the antenna radiation pattern. So the voltage induced in the antenna is given by

$$\hat{V}(\xi, \nu) = \hat{E}(\xi, \nu) \otimes \hat{G}(\xi, \nu)$$

$$= \int_{-\infty}^{\infty} \hat{E}(\xi', \nu)\, \hat{G}(\xi - \xi', \nu)\, d\xi' \qquad 9.19$$

Since the radiation pattern of even a narrow beam antenna departs from a delta function this convolution results in a smearing of the fine structure of the source as indicated in Figure 9.4 (*see* next page). It may be noted that for a point source and for an ideal antenna (zero bandwidth, no sidelobes) we have

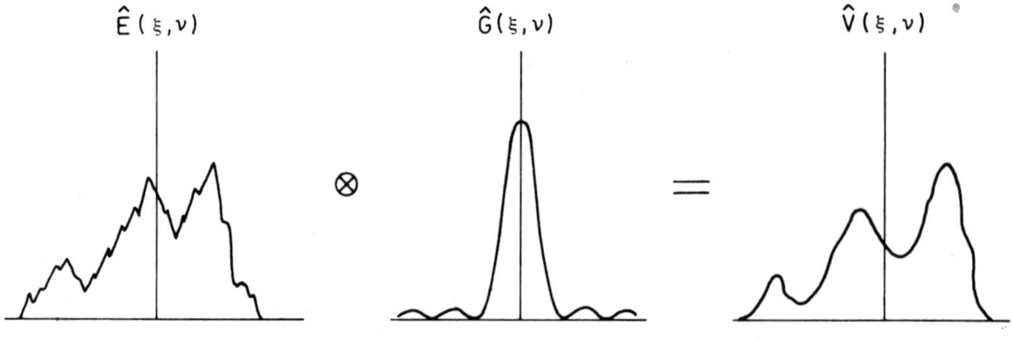

Figure 9.4 Aerial smoothing of the fine structure of the source

$$\hat{V}(\xi, \nu) = \hat{G}(\xi, \nu)$$

$$\hat{V}(\xi, \nu) = \hat{E}(\xi, \nu)$$

respectively. The contamination due to the front end of the receiver is due to its frequency response. If H(t) is the impulse response of the front end its transfer function $\hat{\mathrm{H}}(\nu)$ is the FT of H(t). In the temporal frequency domain the signal input in the correlator is given by

$$\hat{V}_c(\xi, \nu) = \hat{V}(\xi, \nu) \cdot \hat{H}(\nu)$$

$$V_c(\xi, t) = V(\xi, t) \otimes H(t) \qquad 9.20$$

The output of the correlator when two signals are fed to it is given by

$$V_0(\xi, t) = \lim_{T \to \infty} \left[\frac{1}{2T} \int_{-T}^{T} V_{c_1}(\xi, t)\, V_{c_2}^*(\xi, t - \tau)\, dt \right] \qquad 9.21$$

In the light of 9.19-20 and FT relationships we have

$$V_{c_1}(\xi, t) = \int_{-\infty}^{\infty} \left[\hat{H}_1(\nu) \int_{-\infty}^{\infty} \hat{E}(\xi', \nu)\, \hat{G}_1(\xi - \xi', \nu)\, d\xi' \right] e^{j2\pi\nu t}\, d\nu$$

or

$$V_{c_2}(\xi, t) = \int_{-\infty}^{\infty} \left[\hat{H}_2(\nu) \int_{-\infty}^{\infty} \hat{E}(\xi', \nu)\, \hat{G}_2(\xi - \xi', \nu)\, d\xi' \right] e^{j2\pi t\nu}\, d\nu \qquad 9.22$$

We can combine eq 9.21-22 to get the generalized response of the correlator interferometer. By combining all integrations together by using different *dummy* variables of integration the resulting expression for the response is

$$V_0(\tau) = \lim_{T \to \infty} \frac{1}{2T} \left[\begin{array}{l} \int_{-T}^{T} \int\int_{-\infty}^{\infty} \int\int_{-\infty}^{\infty} \hat{H}_1(\nu_1)\, \hat{H}_2^*(\nu_2)\, \hat{E}(\xi_1', \nu_1) \times \hat{E}^*(\xi_2', \nu_2)\, \hat{G}_1(\xi_1 - \xi_1', \nu_1) \\ \times\, \hat{G}_2^*(\xi_2 - \xi_2', \nu_2)\, e^{j2\pi\nu_2\tau} \times e^{j2\pi(\nu_1 - \nu_2)t}\, d\nu_1\, d\nu_2\, d\xi_1'\, d\xi_2'\, dt \end{array} \right]$$

9.23

This formidable expression is considerably simplified by noting the following:

(i) It is well known that

$$\int_{-\infty}^{\infty} e^{j2\pi(\nu_1 - \nu_2)t}\, dt = \delta(\nu_1 - \nu_2)$$

(*ii*) A δ-function in the integrand results in sampling at one point.

(*iii*) By definition,

$$\lim_{T \to \infty} \left[\frac{1}{2T} \left\{ \hat{E}(\xi_1', \nu)\, \hat{E}^*(\xi_2', \nu) \right\} \right] = \hat{\Gamma}(\xi_1', \xi_2', \nu)$$

(*iv*) For incoherent sources,

$$\hat{\Gamma}(\xi_1', \xi_2', \nu) = \hat{\Gamma}(\xi_1', \nu)\, \delta(\xi_1' - \xi_2')$$

Thus the expression simplifies to

$$V_0(\tau) = \int_{-\infty}^{\infty} \int_{-\infty}^{\infty} \hat{\Gamma}(\xi', \nu)\, \hat{H}_1(\nu)\, \hat{H}_2^*(\nu) \times \hat{G}_1(\xi' - \xi, \nu)\, \hat{G}_2^*(\xi' - \xi, \nu)\, e^{j 2 \pi \nu \tau}\, d\xi'\, d\nu \qquad 9.24$$

Usually the front end is a narrow band system having a bandwidth $\Delta\nu$ centred at frequency ν_0 where $(\Delta\nu)/\nu_0 << 1$, for the quasi-monochromatic situation. Neither the source mutual coherence function nor the antenna pattern change significantly within the bandwidth $\Delta\nu$. Eq 9.24 can, therefore, be simplified by carrying out the ν-integration. Assuming that $\hat{H}_1(u)$ and $\hat{H}_2(\nu)$ correspond to rectangular bandpass functions and that the two antenna radiation patterns are identical we get

$$V_0(\tau) = \int_{-\infty}^{\infty} \hat{\Gamma}(\xi', \nu_0)\, |G(\xi' - \xi, \nu_0)|^2\, \frac{\sin(\pi \Delta\nu \tau)}{\pi \Delta\nu \tau}\, e^{j 2 \pi \nu_0 \tau}\, d\xi' \qquad 9.25$$

9.2.2.2 Fringe visibility–brightness distribution relationship

Eq 9.25 is the general expression for the correlator output. While the output appears as a function of τ (the time delay of eq 9.21) it is really a function of the source position relative to the baseline (the line joining the two antennas). The time delay between V_{c1} and V_{c2} is caused by two factors, *viz* (*i*) the geometrical delay τ_g due to source position relative to the baseline, given by $(d \sin \xi_0)/c$, where ξ_0 is the source direction, and (*ii*) the instrumental delay τ_i due to unequal path lengths from the two antennas to the correlator. Consider a small source centred at ξ_0 so that $\Gamma(\xi', \nu_0)$ is nonzero only over a small range of ξ' in the neighbourhood of ξ_0. The instrumental delay is adjusted to cancel the geometrical delay at the source centre, that is $\tau_i = (d \sin \xi_0)/c$. Then at an angle $\xi' = \xi_0 + \theta$, where θ is small, we have

$$\tau(\xi') \approx \frac{d\cos\xi_0}{c}\,\theta = \frac{u\,\theta}{\nu_0} \qquad 9.26$$

Here we have defined $(d\cos\xi_0)/\lambda_0$, which is the baseline normal to the source direction in wavelengths, as the spatial frequency u (in cycles per radian). Then 9.25 can be written as

$$V_0 = \int_{-\infty}^{\infty} \hat{\Gamma}(\xi_0+\theta,\,\nu_0)\,|\hat{G}(\theta,\nu_0)|^2 \left[\frac{\sin\left(\pi\dfrac{\Delta\nu}{\nu_0}u\,\theta\right)}{\pi\dfrac{\Delta\nu}{\nu_0}u\,\theta}\right] e^{j2\pi u\theta}\,d\theta \qquad 9.27$$

In the quasi-monochromatic case the bandwidth pattern, that is the factor within the square brackets in the integrand, is very wide compared to the source size. Also the antenna pattern is much wider than the source size. Ignoring their dependence on θ, the basic relationship between the correlator output and source brightness distribution is given by

$$V_0(u) = \int_{-\infty}^{\infty} \hat{\Gamma}(\theta)\,e^{j2\pi u\theta}\,d\theta \qquad 9.28$$

Eq 9.28 is a Fourier Transform relationship and shows that the correlator interferometer measures a component of the spatial frequency spectrum of the source distribution. This, from eq 9.11 and 9.18, is the fringe visibility. The fringe visibility, a directly measurable complex quantity, can be used to recover the source brightness distribution by inverse FT. This needs measurement of $V_0(u)$ for a whole range of u. This can be achieved by having a variable baseline interferometer or having a large number of interferometers with different baselines. The limitation of this technique is brought out by eq 9.27. A large bandwidth essentially narrows down the bandwidth pattern and acts as a limitation on the field of view, the same effect that a narrow antenna pattern has.

9.3 Aperture synthesis

In general, the two-dimensional brightness distribution $\Gamma(\theta,\phi)$ can be obtained from the measured fringe visibility through the relation

$$\Gamma(\theta,\phi) = \int_{-\infty}^{\infty}\int_{-\infty}^{\infty} V_0(u,v)\,e^{j2\pi(u\theta+v\phi)}\,du\,dv \qquad 9.29$$

In effect, by measuring $V_0(u,v)$ up to sufficiently large baselines one synthesizes an aperture of the size of the largest baseline. To fill this synthesized aperture completely, all intermediate values of u and v must be realized. The spatial frequencies u, v correspond to the components of the projection of the baseline on a plane normal to the direction of the source.

9.3.1 Supersynthesis

Since the apparent source position changes as the earth rotates a single antenna pair can measure the fringe visibility $V(u,v)$ for a whole range of values of u,v. This method of using earth rotation to measure a large number of Fourier components is called *supersynthesis*. It was first proposed by Sir Martin Ryle of the University of Cambridge for which he was awarded the Nobel Prize (Ryle 1975). This has revolutionized radio telescope design and today all large instruments all over the world are using this technique to obtain high resolution maps of extended radio sources.

9.3.2 Sampling and field of view

Eq 9.27 shows that if we consider a point source in the quasi-monochromatic situation the fringe visibility measurements give the radiation pattern of the synthesized antenna. For convenience of measurement as well as Fourier transformation through the use of *fast Fourier transformation* (*FFT*), one measures $V_0(u,v)$ at discrete intervals. The result of this is that the synthesized antenna beam will show grating lobes, *i. e.* will be periodic in θ, ϕ. For unambiguous mapping the grating lobes must fall outside the source extent. Thus if the maximum source size is θ_{max}, ϕ_{max} the sampling intervals Δu and Δv should be chosen so that

$$\Delta u \leq \frac{1}{\theta_{max}}, \; \Delta v \leq \frac{1}{\phi_{max}} \tag{9.30}$$

In practice it is quite adequate to let the grating lobes be outside the field of view defined by the antenna element pattern or bandwidth pattern (whichever is narrower).

Similarly, if one needs to map the source with a resolution θ_{min} , ϕ_{min} it is necessary to measure visibilities up to baselines corresponding to

$$u_{max} = \frac{1}{\theta_{min}}, \; v_{max} = \frac{1}{\phi_{min}} \tag{9.31}$$

9.3.3 Transfer function and beam pattern

In a correlator array each antenna pair produces an output which gives one Fourier component corresponding to the projected baseline. Thus for a given distribution of antennas in the x,y-plane the instantaneous synthesized aperture consists of a number of points corresponding to the autocorrelation function of the antenna distribution in the u,v-plane. This is called the transfer function of the array. An example of a five-element

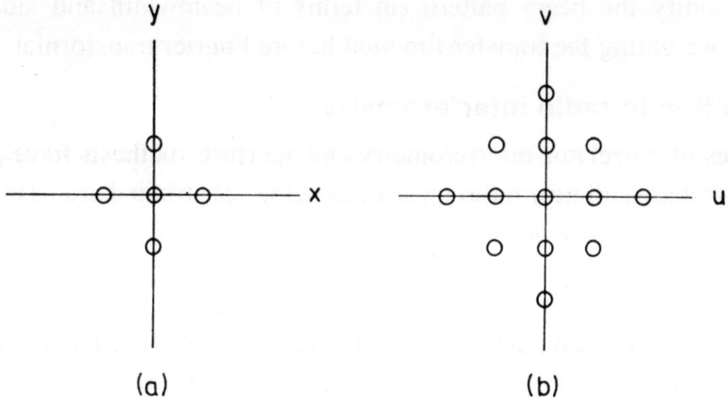

Figure 9.5 Transfer function of a correlator array, (a) Aperture illumination, (b) Transfer function

array is shown in Figure 9.5a. In the supersynthesis mode each baseline measures a number of Fourier components. It can easily be shown that as the earth rotates each antenna pair measures Fourier components lying along an elliptical track in the u, v-plane. For a baseline of length d, if the projected baseline corresponds to a point on the celestial sphere having declination δ_B and hour angle H_B, and the source being observed is at location δ_s, H_s, we have

$$u = \frac{d}{\lambda} \cos\delta_B \sin(H_s - H_B) \qquad 9.32$$

$$v = \frac{d}{\lambda}[\sin\delta_B \cos\delta_s - \cos\delta_B \sin\delta_s \cos(H_s - H_B)]$$

As the earth rotates, the source hour angle H_s varies from 0 to 24 hours (from 0 to 360°). Eliminating the time variation in eq 9.32 we get

$$\left[\frac{u}{d\cos\delta_B/\lambda}\right]^2 + \left[\frac{v - d\sin\delta_B \cos\delta_s/\lambda}{d\cos\delta_B \sin\delta_s/\lambda}\right]^2 = 1 \qquad 9.33$$

which is the equation of an ellipse in the u, v-plane. Since the u, v-plane is divided into a grid for beam pattern calculations each grid cell is taken to be sampled if one of the elliptical tracks passes through it. Thus the design of a correlator supersynthesis array consists in optimizing antenna locations so that the largest number of cells are sampled within the overall synthesized aperture (Māthur 1969). The beam pattern is obtained by taking the FT of the transfer function. There are two features of this beam pattern of a synthesized aperture which are different from those of an ordinary aperture antenna. First, the beam pattern is different in different directions since it depends upon the source declination. It also depends upon the length of time for which the source is observed. Second, it is possible to modify the beam pattern (in terms of beamwidth and sidelobe levels) by appropriately weighting the transfer function before Fourier transformation.

9.4 Application to radio interferometry

The techniques of correlator interferometry and aperture synthesis have proved so highly successful that almost all new radio telescopes being constructed are based on them. There are two main areas of application.

9.4.1 High resolution mapping

As discussed in the preceding section, a correlator array enables one to map the brightness distribution of an extended source by measuring a number of Fourier components of the spatial frequency spectrum of this brightness distribution and then carrying out a Fourier inversion. The resolution obtainable by this technique is limited only by the longest baselines which can be achieved. The quality of the synthesized beam depends upon the transfer function of the array which, in turn, depends upon the number and size of the apertures, their mobility and steerability, and the data processing technique. A large number N of antennas judiciously located enables one to cover a large number of independent elliptical tracks [up to N(N−1)/2]. The synthesized beam is usually poorest for equatorial sources (that is, declination is zero) since the elliptical degenerates into a straight line for this declination. Beam quality can be controlled by weighting the transfer function appropriately to suppress large sidelobes. Sophisticated techniques such as maximum entropy method and *CLEAN* have been developed for removing the contaminating effects of a non-ideal beam on the measured map (Burg 1979, Hogborn 1974). They are deconvolution techniques which essentially use *a priori* information about the source structure. There are also difficulties in the accurate determination of the phase of the visibility due to atmospheric effects. While it was earlier felt that the atmospheric phase fluctuation put the ultimate limit on high resolution mapping new techniques such as the closure phase technique have been developed to get around this difficulty (Cornwell and Wilkinson 1981).

Very many radio telescopes using the aperture synthesis principle have been constructed. Table 9.1 summarizes some data about them. The largest and most sophisticated of them is the *Very Large array (VLA)* in new Mexico, USA. It consists of 27 steerable antennas, each 25 m in diameter, which are movable on rail tracks laid in the form of a *Y* having arms of length 21 Km (Thompson *et al* 1980, Hjellming and Bignell 1982). It can operate at frequencies of 1.4, 5, 14 and 23 GHz and has recently been modified to operate also at 327 MHz. At the highest frequency the synthesized beam is approximately 0.17×0.25 arcsec.

9.4.2 Source size and position measurement and Very long baseline interferometry

It has been pointed out earlier that a radio source will produce interferometric fringes only if the source size is smaller than the fringe spacing. Thus a two-element interferometer can be used to place an upper limit on the source size. Also, by accurate measurement of the visibility phase, the exact source position can be determined (provided, of course, that *a priori* information is available on the approximate position).

In ordinary, connected-antenna interferometers the resolution is limited by the largest antenna spacing which will permit appropriate interconnection. This limits the baseline to a few tens of kilometres (*e. g.* the longest baseline of VLA is about 35 Km). In 1967 a new technique called *Very Long Baseline Interferometry (VLBI)* was first implemented which removed this limit on baseline length. In this system (*see* Readhead 1982, for example) the two antennas observe the source simultaneously but the signals are recorded independently at the two sites on magnetic tape. The signals are later crosscorrelated in a computer. Along with the signals (which are recorded in bipolar form after sampling) accurate information about time is also recorded in order to obtain proper delays for correlation. Due to the long baselines the fringe rate is very high and extremely stable local oscillators (derived from rubidium vapour frequency standards or hydrogen masers) have to be used. VLBI permits the use of extremely long baselines up to several thousand kilometres (in fact the ultimate limit is the diameter of the earth), the only consideration being that the source should be simultaneously observable from the two sites. With VLBI many of the smaller sources which were earlier considered to be point sources have been resolved and detailed structures have been seen. Resolutions better than a milli-arcsec have been achieved. The technique not only enables accurate position measurements to be made but also has applications in such diverse fields as measurement of plate motions and continental drift, measurement of gravitational bending of light and confirmation of Einstein's theory, measurement of polar motion, synchronization of clocks, *etc*. Of late substantial work has been done in successfully operating *Very Long Baseline Arrays (VLBA)* and obtaining high resolution maps by aperture synthesis. Construction started in May 1985 for a VLBA which will use 25 m antennas in locations from Hawaii to eastern USA and operate at frequencies up to 45 GHz (Kellerman 1985).

9.5 The Indian scene

Radio astronomy is one field in which India has kept pace with the developed countries, thanks to the early efforts started at the Tata Institute of Fundamental Research, Bombay, in 1964. The giant Ooty radio telescope has become part of a large synthesis telescope which currently operates at 327 MHz. Another giant facility operating at metre wave-

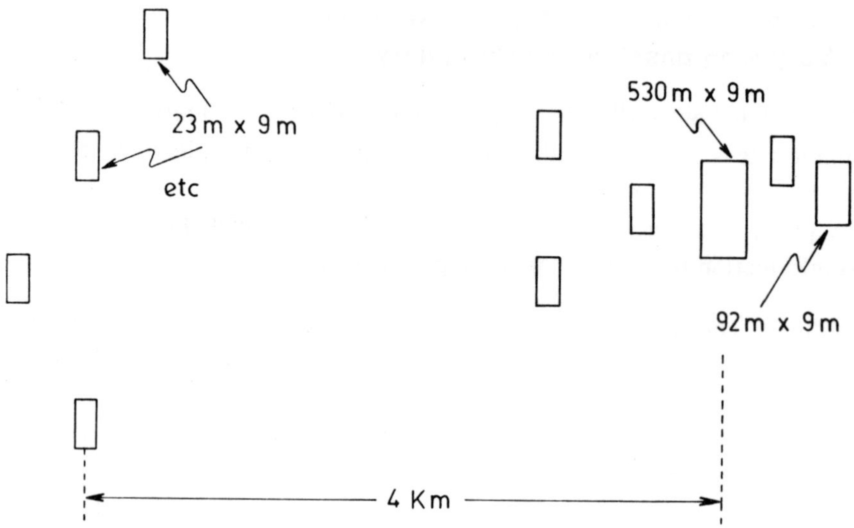

Figure 9.6 The Ooty Synthesis Radio Telescope

lengths has been approved and funded for construction near Puné. Brief descriptions of these systems are given below.

9.5.1 Ooty Synthesis Radio Telescope (OSRT)

The OSRT has been constructed around the 530m × 30m Ooty Radio Telescope *(ORT)*. It operates at 327 MHz and currently has 10 antennas located as shown in Figure 9.6. The ORT is divided into five parts and works as five independent smaller antennas. Each antenna consists of a steerable rectangular parabolic cylinder capable of tracking a source from horizon to horizon. Because of the hilly terrain antennas are located at accessible positions only and the antenna configuration cannot be optimized for the best u, v-coverage. The system has a declination coverage of $-40°$ to $+40°$ and has a resolution of 40 × 80 arcsec at declination zero. Because of terrain difficulties many of the antenna pairs are connected by microwave links for local oscillator and IF-signals.

9.5.2 Giant Metre-wavelength Radio Telescope (GMRT)

The GMRT will be the largest synthesis radio telescope in the world planned for operation at metre wavelengths (Swarup 1984). It will consist of 34 parabolic cylindrical antennas each 92m × 35m arranged as shown in Figure 9.7 (*see* p. 135). They will be steerable from $-30°$ to $+55°$ in declination and horizon-to-horizon in hour angle. The system will operate at frequencies of 38, 152, 326 and 610 MHz and will provide a remarkable resolution of 75

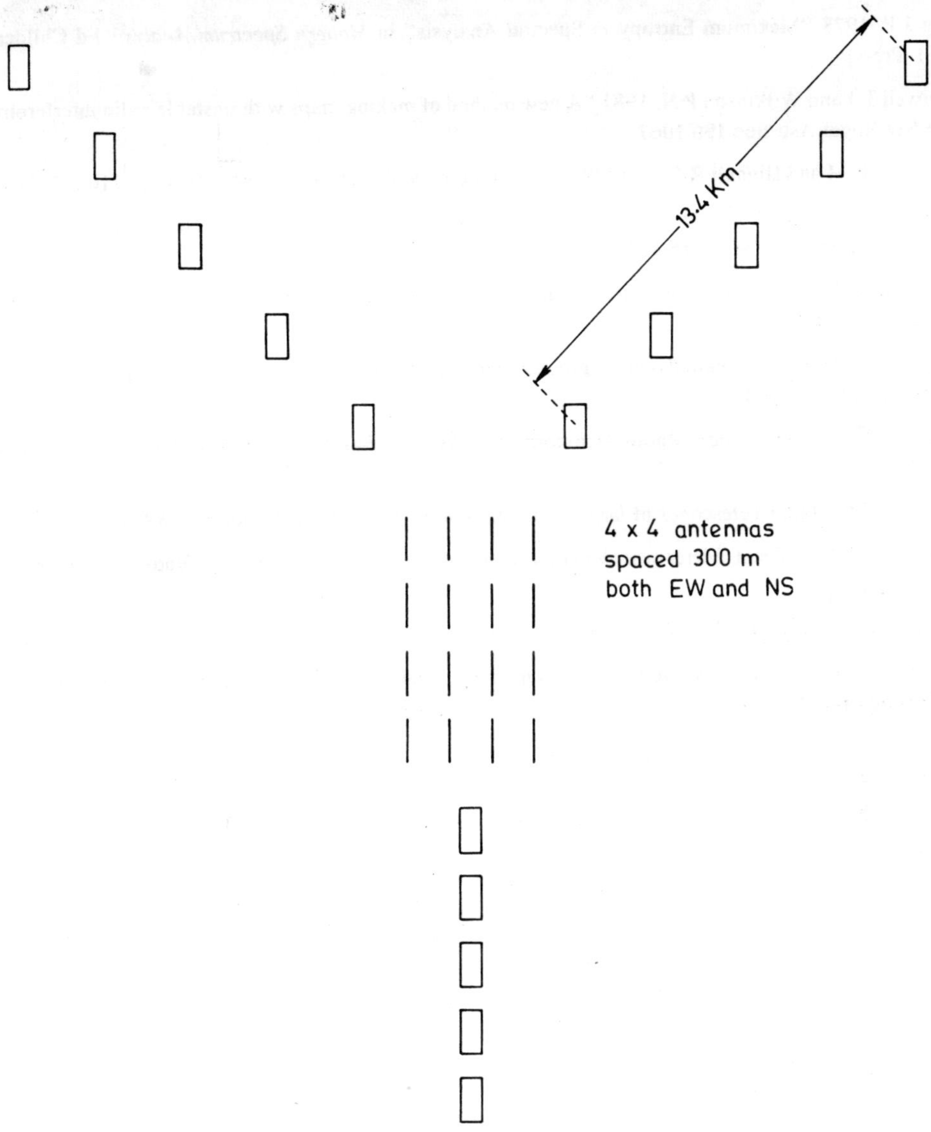

Figure 9.7 Giant Metre-Wavelength Telescope

arcsec even at 38 MHz. The ionospheric phase distortions are proposed to be taken care of by the techniques mentioned earlier.

Besides providing detailed maps of galactic and extra-galactic radio sources the GMRT will also be used in a global VLBA system.

References

Born M and Wolf E, 1975 *Principles of Optics* 5th ed (London: Pergamon Press)

Burg J P, 1978 "Maximum Entropy in Spectral Analysis" in *Modern Spectrum Analysis* Ed Childers D G (IEEE Press)

Cornwell T J and Wilkinson P N, 1981 "A new method of making maps with unstable radio interferometers" Mon Not Royal Astr Soc **196** 1067

Hjellming R M and Bignell R C, 1982 "Radio astronomy with very large array" Science **216** (4552) 1279

Hogborn J A, 1974 "Aperture synthesis with a nonregular distribution of interferometer baselines" Astronomy and Astrophysics Suppl **15** 417

Kellerman K I, 1985 "Very long baseline arrays" in *XIX General Assembly of International Astronomical Union*, New Delhi

Māthur N C, 1969 "A pseudodynamic programming technique for the design of correlator supersynthesis arrays" Radio Science **4** 235

Readhead A C S, 1982 (June) "Radio Astronomy by Very Long Baseline Interferometry" Scientific American **246** 38

Ryle M, 1975 "Radio Telescopes of large resolving power" Nobel Lecture Science **188** 1071

Swarup G, 1984 "Giant Meter-wavelength Radio Telescope — A Proposal" Radio Astronomy Centre Ootacamund

Swenson G W (Jr) and Māthur N C, 1968 "The interferometer in Radio Astronomy" Proc IEEE **56** (12) 2114

Thompson A R, Clark B G, Wade C M and Napier P J, 1980 "The Very Large Array" The Astrophysical J Suppl Series **44** 151

Chapter 10

X-RAY IMAGING

H C PANT

10.1 Introduction

In x-ray optics there are no new ideas, only new technologies to exploit. The years 1975-85 brought a wide array of advances in x-ray optics placing it on the threshold of achievements which could significantly transform fields like microbiology and material microanalyses. Several important contributions to other areas like plasma physics, surface physics and atomic physics are also expected. This exciting situation results not from any conceptual break-through or new understanding of x-ray optical principles. It results instead from advances in technology which for the first time make possible the precision implementation of concepts which are several decades old.

A number of technologies have contributed to the rapid advances in x-ray optical capabilities. Improvements in the ability accurately to figure, polish and measure reflective surfaces have led to x-ray microscopes and telescopes of unprecedented resolution and reflection efficiency. The availability of synchrotron and storage ring facilities as well as high power lasers, particle beam and plasma discharge devices now provide a variety of bright x-ray sources with varied spectral, temporal, collimation and coherence characteristics. Advances in solid state technology and photoresist chemistry have provided high resolution methods of significantly improved sensitivity.

10.2 X-ray pinhole camera

A pinhole camera is an optical imaging device which consists of a small circular pinhole aperture in an opaque screen. Each point of a luminous object casts a shadow of the screen through the clear pinhole aperture on a viewing screen. Thus the image of the luminous object is a shadowgraph and the process of imaging *shadowgraphy*. Imaging or shadowgraphy by a pinhole camera is in contrast to imaging by other optical imaging devices or components in which the images are formed by either refraction or reflection. The pinhole camera is perhaps the first known optical instrument other than the looking glass. Although

De Porta (c 1600) is believed to be the inventor of the pinhole camera it was used earlier by Leonardo da Vinci and was perhaps known to the ancient Greeks too. In spite of its antiquity and apparent simplicity it has a number of advantages over more sophisticated optics. In applications where resolution is not a major factor it offers an inexpensive useful instrument, free from distortion, having a good depth of field and a wide angular field, almost 90°. However, pinhole cameras have been used to obtain 1 μm spatial resolution with radiation of wavelength of scores of nanometres. With fast films and recorders pinholes of low aperture (*e. g.* f/1000) are no more a limitation.

Among the many applications of the pinhole camera may be mentioned that of a pinhole television camera which was used for a flight simulator that required both a large depth of field and a large angular field. The choice of pinhole optics over conventional automatic focussing systems was made to attain clarity from 7.6 cm to 107.6 m! X-ray lithography which uses a pinhole array 8 × 8 camera has been utilized in manufacturing integrated circuits.

The x-ray pinhole camera is being used extensively in space-resolved measurements of x-rays emitted by very small hot sources like those produced during laser fusion experiments.

In order to establish a relationship between the pinhole aperture size and the geometrical and wavelength parameters which will give the best resolution of the image it is essential to discuss the limits of resolution in the geometrical ($a >> \lambda$) and wave ($a \sim \lambda$) approximations, a and λ being the aperture diameter and x-ray wavelength respectively. The image formation by a pinhole camera in the geometrical or ray approximation is independent of the radiation wavelength whereas in the wave approximation it is wavelength-dependent.

10.3 Geometrical blurring

A point object casts a shadow of the opaque plane screen containing the pinhole aperture. This forms a blur spot on the viewing screen (Figure 10.1a, next page). The dimension of this blur spot depends on the pinhole aperture diameter and the magnification of the system involving the geometrical parameters, *i. e.* the object and image distances from the pinhole aperture. The geometrical blur spot diameter d_G is found thus:

$$\frac{d_G}{a} = \frac{p+q}{p} \equiv 1 + \mu$$

$$d_G = a(1 + \mu) \tag{10.1}$$

where p and q are the object and image distances respectively and μ is called the *magnification factor*.

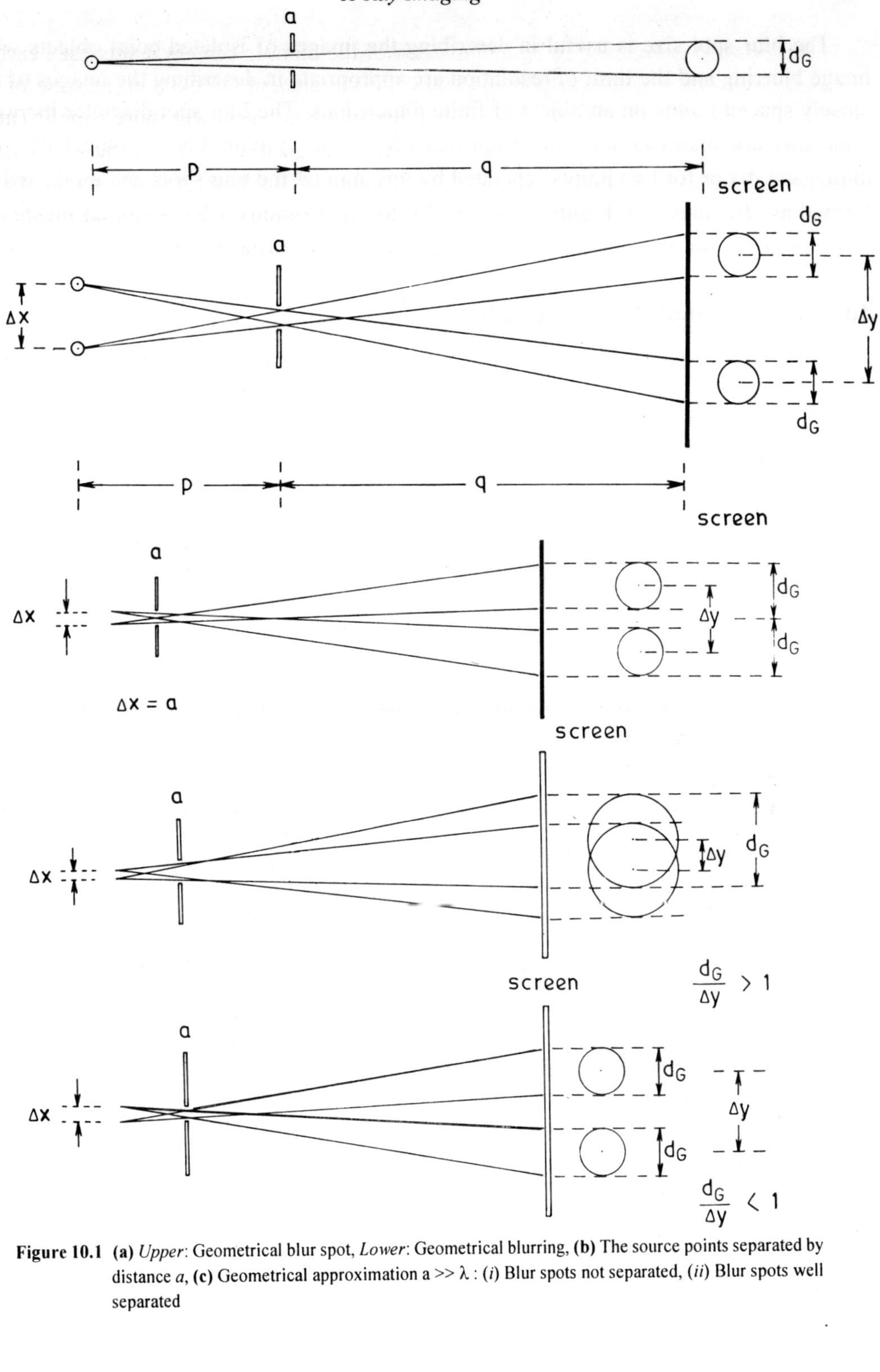

Figure 10.1 **(a)** *Upper*: Geometrical blur spot, *Lower*: Geometrical blurring, **(b)** The source points separated by distance *a*, **(c)** Geometrical approximation a >> λ : (*i*) Blur spots not separated, (*ii*) Blur spots well separated

The blur spot size is useful in describing the images of isolated point objects while image blurring and the limit of resolution are appropriate in describing the images of two closely spaced points on an object of finite dimensions. The blur spot diameter increases with aperture diameter and the magnification of the system. For an object of finite dimension Δx or for two points separated by this amount the blur spots and image will be formed as also shown in Figure 10.1a (p 139, lower diagram). Blurring of an image of a finite object is governed by the blur spot size d_G and separation of blur or image spots Δy.

10.4 Limit of resolution (Ray optics)

The geometrical blurring of an image is defined as the ratio of blur spot size d_G to image size Δy. We have

$$\frac{d_G}{\Delta y} = \frac{[a(1+\mu)]}{\Delta y}$$

$$\mu \equiv \frac{q}{p} = \frac{\Delta y}{\Delta x}$$

$$\frac{d_G}{\Delta y} = \frac{a}{\Delta x}\left(1 + \frac{1}{\mu}\right) \qquad 10.2$$

Thus for a given object size Δx or two point objects separated by this amount the blurring increases with aperture dimension but decreases with magnification. From eq 10.1 and 10.2 it is obvious that both blur spot size and blurring of the image increase with aperture size. On the other hand, though the blur spot size increases with μ the image blurring decreases with μ. Let us now consider three limiting cases of eq 10.2:

$$\frac{d_G}{\Delta y} = 1, > 1, < 1$$

(*i*) *Blurring* ($d_G / \Delta y = 1$): This implies that the size of the image spots or blur spots of two point objects separated by distance Δx is equal to the image size Δy. In other words, the image spots of two point objects separated by Δx are just resolved. We have then

$$\frac{a}{\Delta x}\left(1 + \frac{1}{u}\right) = 1$$

$$\left[\frac{a}{\Delta x}\left(1 + \frac{1}{\mu}\right)\right]_{\mu >> 1} \approx \frac{a}{\Delta x} \approx 1$$

$$a \approx \Delta x \qquad 10.3$$

Thus for large magnification and an aperture size equal to Δx the image spots are just resolved. This represents the geometrical limit of resolution. The separation of point objects (Δx) and separation of the image spots (Δy) represent the limit of resolution in the object and image spaces respectively.

The limit of resolution is defined in the object space as the closest separation (Δx) of two point objects whose image spots are just resolved.

Thus, from the condition $d_G / \Delta y = 1$, the geometrical limit of resolution is given by

$$[\Delta x]_G = a\left(1+\frac{1}{\mu}\right) \qquad 10.4$$

(*ii*) *Blurring* ($d_G / \Delta y > 1$): This implies that the image spots of two points separated by Δx are larger than the separation between the image spots Δy. From such an image it is not possible to decipher the details of the object illumination. In this case

$$\frac{a}{\Delta x}\left(1+\frac{1}{\mu}\right) > 1$$

and, for large μ, $a > \Delta x$ 10.5

Thus, for large magnification and large pinhole aperture compared to an object size, one would get a uniform illumination on the viewing screen. Because of this particular inequality (*viz* that preceding eq 10.5) we have

$$[\Delta x] < a\left(1+\frac{1}{\mu}\right)$$

$$[\Delta x] < [\Delta x]_G, \quad \text{for } a > \Delta x$$

This means that the image spots of two point objects with separation Δx less than that required for the geometrical limit of resolution $[\Delta x]_G$ will not be resolved in the image space.

(*iii*) *Blurring* ($d_G / \Delta y < 1$): This condition says that the size of the image spots of two point objects is smaller than their separation Δy. The two image spots are therefore distinctly resolved. However, the resolution will be the poorer in this case. We have

$$\frac{a}{\Delta x}\left(1+\frac{1}{\mu}\right) < 1$$

and, for large μ, $a < \Delta x$

Thus, for large magnification μ and a pinhole aperture size smaller than the separation between two object points Δx, the image spots are well resolved. Here we have $[\Delta x] > [\Delta x]_G$ for $a < \Delta x$. From the inequality it is clear that the image spots of two point objects with a separation greater than that required for geometrical limit of resolution will be resolved distinctly. However, the resolution will be the poorer since only the points in the object space separated by a distance greater than that required for the geometrical limit of resolution will be resolved distinctly. Thus it will not be possible to decipher better details of an object from its image than those in the case of the limit of resolution.

We conclude that in the geometric limit the resolution improves with a decrease in pinhole aperture diameter. Blur spot size and blurring also decrease with decrease in the pinhole diameter. It would appear that the blur spot size of a point object can be reduced indefinitely to point image by reducing the pinhole diameter.

10.5 Wave approximation (diffraction limit)

In fact, however, a continuous decrease in aperture size leads to a stage where aperture dimensions become comparable to the wavelength of the object radiation and the ray or geometrical approximation breaks down. For small apertures physical or wave optics dominates and the image is formed by the diffraction phenomenon.

In the wave approximation the image of a point source at infinity is no longer a blur spot but an Airy pattern (*see* § 2.8.3). At finite but long enough distances the pattern is at least approximately Airy pattern. Thus a point source is not imaged as a point image. The central bright spot of the Airy or quasi-Airy pattern carries the Airy disc, that is, the central bright spot in an Airy pattern may be called the *diffraction blur spot*; it plays the same role in image blurring and image resolution in the wave approximation as the geometrical blur spot plays in the geometrical approximation.

The irradiance distribution is given in terms of a normalized Bessel function of the first order (Figure 10.2, next page) as

$$I = \left[\frac{2\,J_1(Z)}{Z}\right]^2$$

$$\lim_{Z \to 0} \left[\frac{J_1(Z)}{Z}\right] = \frac{1}{2}$$

$$Z = \frac{k\,a\,(\sin\alpha)}{2}$$

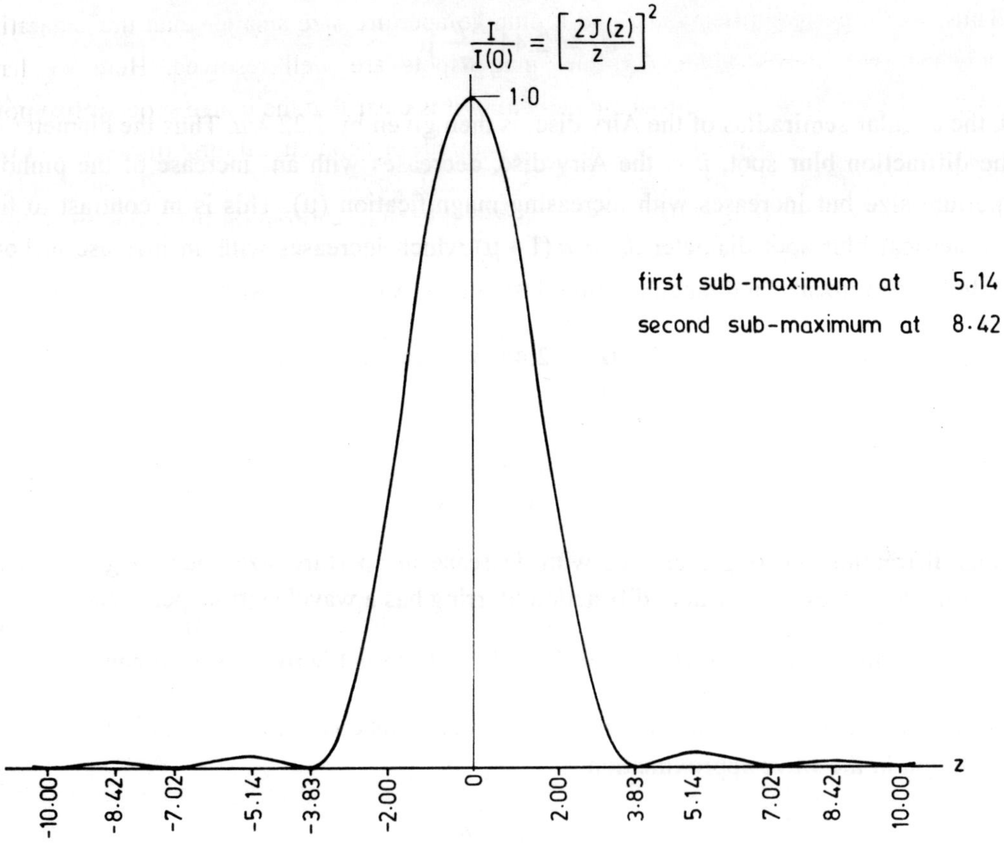

Figure 10.2 Intensity distribution across the Airy pattern

where $a/2$ is the radius of the aperture and α the semiangle of the emergent cone. As α tends to zero, Z also does, and $J_1(Z)/Z$ tends to ½, we get I = 1. The radius $d_A/2$ and diameter d_A of the central bright spot or the Airy disc are

$$\frac{d_A}{2} = \frac{1.22\,\pi}{k\sin\alpha} = \frac{0.61\,\lambda}{\sin\alpha}$$

$$d_A = \frac{1.22\,\lambda}{\sin\alpha}$$

Putting in $\sin\alpha = a/2q$ and $\mu = q/p = \Delta y/\Delta x$ we get the radius and diameter as

$$\frac{d_A}{2} = 1.22\,\frac{\lambda p}{a}\,\mu$$

$$d_A = 2.44 \frac{\lambda p}{a} \mu$$

θ, the angular semiradius of the Airy disc, is then given by 1.22 λ/a. Thus the diameter of the diffraction blur spot, *i. e.* the Airy disc, decreases with an increase of the pinhole aperture size but increases with increasing magnification (μ). This is in contrast to the geometrical blur spot diameter $d_G = a(1 + \mu)$ which increases with an increase in both pinhole aperture size and magnification. Using $\mu = \Delta y / \Delta x$ we write

$$\frac{d_A}{\Delta y} = \frac{2.44}{\Delta x} \frac{\lambda}{a} p$$

$$= \frac{2.44}{\Delta x} \frac{\lambda}{a} \frac{q}{\mu}$$

Thus diffraction blurring increases with decrease in aperture size whereas geometrical blurring decreases. In addition, diffraction blurring has a wavelength-dependence.

Now consider the cases $d_A / \Delta y = 1, > 1, < 1$. (*See* Figure 10.3, next page.)

(a) $d_A / \Delta y = 1$: The images are just resolved. This condition provides in the limit of resolution in the wave approximation

$$[\Delta x]_D = 2.44 \left(\frac{\lambda}{a}\right) p$$

$$= 2.44 \left(\frac{\lambda}{a}\right)\left(\frac{q}{\mu}\right)$$

If μ is large, the resolution is small, while in the geometrical limit we had

$$[\Delta x]_G = a(1 + \mu^{-1}) \approx a$$

(b) $d_A / \Delta y > 1$: The diffraction blur spots merge into each other and the details of the object on the image plane are lost. The condition implies $\Delta x < [\Delta x]_D$, which shows that the points in the object space separated by Δx smaller than $[\Delta x]_D$ will not be resolved in the image.

(c) $d_A / \Delta y < 1$: Then we have

$$\Delta x > 2.44 \left(\frac{\lambda}{a}\right) p$$

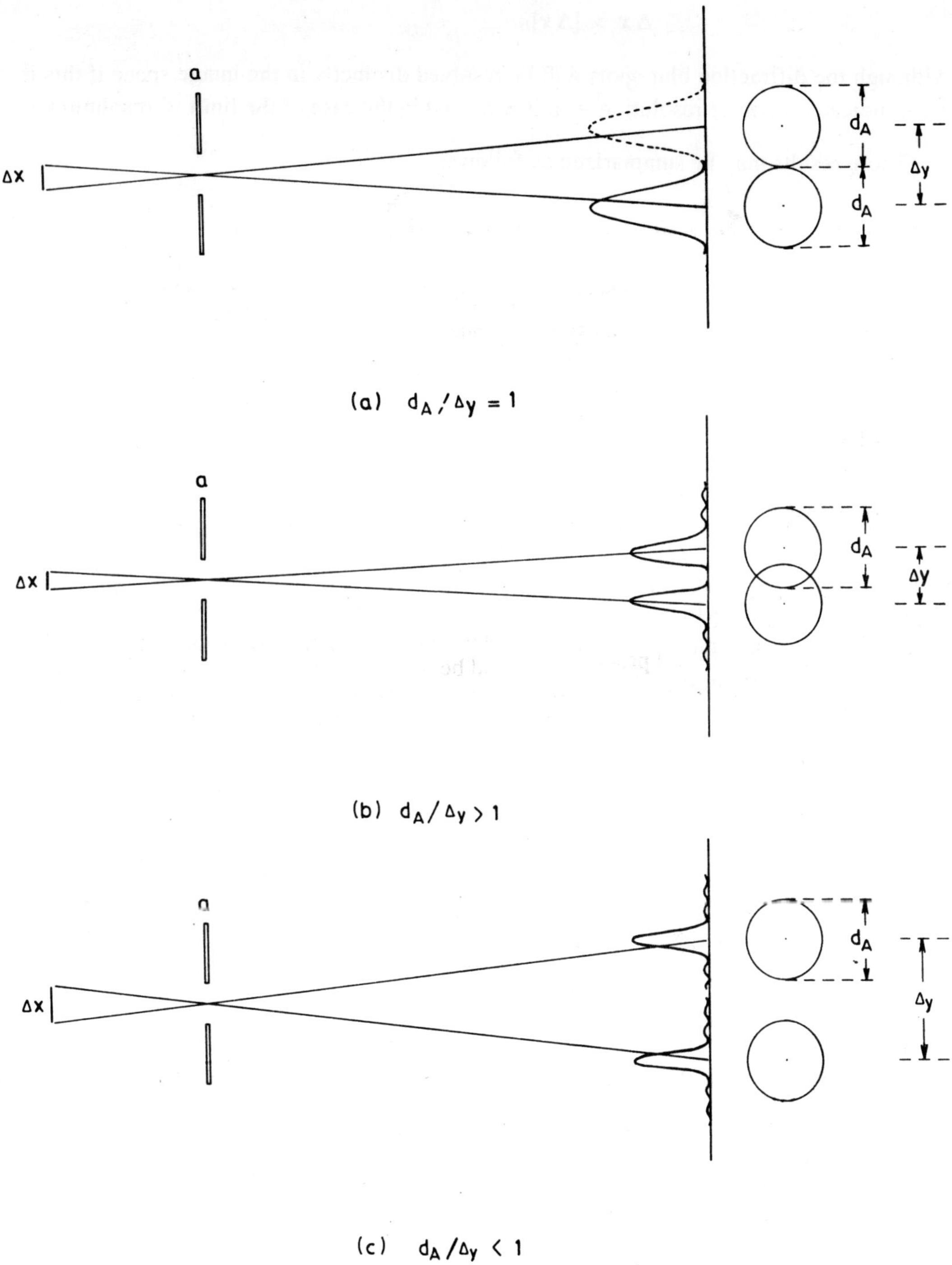

Figure 10.3 Resolution in terms of the wave approximation: $d_A / \Delta y$: (*a*) = 1, (*b*) > 1, (*c*) < 1 [In (b) and (c) the Airy disc is not shown to scale with the irradiance curve].

$$\Delta x > [\Delta x]_D$$

Although the diffraction blur spots will be resolved distinctly in the image space if this is so, it indicates a poorer resolution than is achieved in the case of the limit of resolution.

These results may be summarized as follows:

Table comparing the results of geometrical and wave optics

	Geometrical optics $(a \gg \lambda)$	**Wave optics** $(a \sim \lambda)$
blur spot size	$d_G = a(1+\mu)$	$d_A = 2.44 \frac{\lambda p}{a} \mu$
image blurring	$\frac{d_G}{\Delta y} = \frac{a}{\Delta x}\left(1 + \frac{1}{\mu}\right)$	$\frac{d_A}{\Delta y} = \frac{2.44\,\lambda}{\Delta x\; a} p$
limit of resolution	$[\Delta x]_G = a\left(1 + \frac{1}{\mu}\right)$	$[\Delta x]_D = 2.44 \frac{\lambda}{a} p$

Thus, in order to achieve a near-perfect image with a good resolution, it is desired to choose a pinhole aperture size which will give the smallest (*a*) blur spot size, (*b*) image blurring and (*c*) limit of resolution.

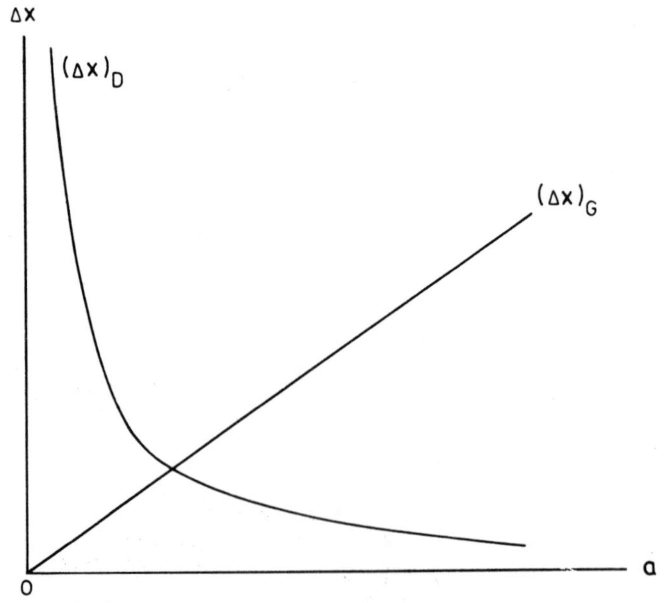

Figure 10.4 Variation of the limit of resolution with aperture size *a* in the two approximations, Geometrical and Wave (Schematic only; ordinate scales for the two different)

For geometrical imaging ($a \gg \lambda$) all quantities decrease with decrease in a but precisely that operation increases them all in the wave approximation ($a \sim \lambda$). The graph of the limit of resolution *versus* pinhole aperture size in the former is a straight line and in the latter a rectangular hyperbola. Neither of the approximations is strictly valid where the two cross. However, the best resolution in both approximations will be given by this region. Thus the optimum pinhole aperture size a_{opt} can be obtained from the intersection of these curves where $[\Delta x]_G = [\Delta x]_D$ (Figure 10.4, p 146). Thus we get

$$a_{opt}\left[1 + \frac{1}{\mu}\right] = 2.44\,\frac{\lambda}{a_{opt}}\,p$$

$$a_{opt} \approx 1.6\left[\frac{\lambda p}{1 + 1/\mu}\right]^{\frac{1}{2}} \qquad 10.6$$

or, for $\mu \gg 1$,

$$a_{opt} \approx 1.6\,(\lambda p)^{\frac{1}{2}}$$

For a pinhole size $a < a_{opt}$ the image will be degraded due to diffraction effects. Thus

$$a \geq a_{opt}$$

$$\geq 1.6\,[\lambda p]^{\frac{1}{2}}$$

$$p \leq \frac{a^2}{2.44\,\lambda}$$

When this condition is satisfied the image is free from diffraction effects. Eq 10.6 can also be written as

$$\frac{1}{p} + \frac{1}{q} = \frac{2.44\,\lambda}{a^2} \qquad 10.7$$

We see that a pinhole camera with a hole size of a will act as a lens of focal length $f = a^2/2.44\lambda$.

10.6 Kirkpatrik-Baez x-ray microscope *(KBXRM)*

This system was first developed by Kirkpatrik and Baez at Stanford in 1948. The principle is the same as using two cylindrical refracting lenses at 90° apart in rotation in visible optics. Here one employs grazing incidence reflection of x-rays from cylindrically polished x-ray mirrors. The x-rays reflecting from one cylindrical mirror produce a line image as do the x-rays reflecting from the other cylindrical mirror. The x-rays that reflect from both of the surfaces produce the double reflection image of the object.

These microscopes are subject to geometrical aberrations which limit the resolution and field of view. The two main aberrations are spherical aberration and obliquity of the field. In experiments the apertures are kept extremely small so as to limit the effects of spherical aberration. However, the extremely small solid angle subtended by the microscope (~ 10^{-7} Sr) makes diffraction also a limiting factor. The best resolution that has been obtained with these microscopes is 2-3 μm.

The geometrical spot size from spherical aberration is

$$S_G \approx \frac{3}{8}\,\varepsilon\,\frac{W^2}{R}\,\frac{\mu - 1}{\mu} \qquad 10.7a$$

and the diffraction spot size is

$$S_D \approx 0.886\,\frac{d_0\,\lambda}{W \sin i} = 1.1 \times 10^{-5}\,\frac{d_0}{W\,i\,E} \qquad 10.8$$

where W is the full mirror width in mm, R the mirror radius of curvature in mm, μ the magnification, d_0 the object distance, *i* the grazing incidence angle in radians, *E* the average photon energy in KeV, λ the average photon wavelength in nm, ε a numerical constant equal to 0.33. The angle of incidence is related to the mirror high energy cutoff and to the average photon energy *E* of the radiation as

$$1 = 0.019\,\sqrt{\rho}\,/\,E_c = 0.019\beta\,\sqrt{\rho}\,/\,E \qquad 10.9$$

where β is the ratio of the channel's average photon energy to the mirror's cutoff energy. Now

$$f = \frac{R}{2}\sin i \approx \frac{R_i}{2} \qquad 10.10$$

and the thin lens formula is

$$\frac{1}{f} = \frac{1}{d_0} + \frac{1}{\mu\,d_0} = \frac{\mu + 1}{\mu\,d_0} \qquad 10.11$$

Assuming that the real spot size is S_0 and using eq 10.9-11 to eliminate R, i and f we get

$$S_0^2 = S_G^2 + S_D^2$$

Optimizing S_0 with respect to W yields $S_D = (\sqrt{2})\, S_G$ at $W = W_0$ which in turn yields

$$S_0 = \sqrt{3}\ S_G = \sqrt{(3/2)}\ S_D = 3.14 \times 10^{-4}\left[\frac{\varepsilon\, d_0}{\beta\sqrt{\rho}\, E}\left(\frac{\mu^2 - 1}{\mu^2}\right)\right]^{\frac{1}{3}} mm \qquad 10.12$$

and

$$W_0 = 0.225\left[\frac{d_0^2\, E}{\varepsilon\, \beta^2\, \rho}\left(\frac{\mu^2}{\mu^2 - 1}\right)\right]^{\frac{1}{3}} mm \qquad 10.13$$

These are the optimum resolution and mirror width for a channel.

The x-ray microscope shown in Figure 10.5 has been built at the Lawrence Livermore National Laboratory, USA. It consists of two polished cylindrical concave mirrors of radius of curvature R ~ 19.7 m. Each element acts as a cylindrical lens for x-rays incident on the surface at an angle θ of nearly 1°. Each element brings rays from a point source to a line focus (f = 17 cm). The two elements are oriented so that a small fraction of the source x-rays is reflected from both surfaces. The planes of reflection are mutually perpendicular so that the combination of the two line foci is a true image of the source. The system has a magnification of 3. The effective solid angle subtended at the target is ~ 6×10^{-8} Sr.

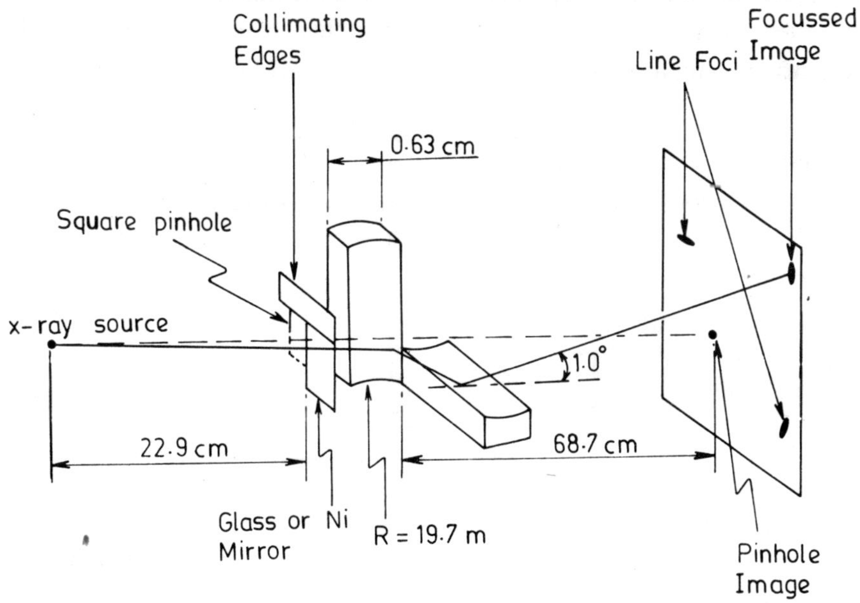

Figure 10.5 Grazing incidence x-ray reflection microscope

Following are the advantages of an x-ray microscope:

(i) Spatial resolution up to 3 μm can be obtained by a single lens system.

(*ii*) The lens can be positioned far from the source.

(*iii*) Focal conditions are not critical in a single lens system.

10.7 Wolter axisymmetric x-ray microscope

The KB-microscope is very versatile and has a large field of view. However, it has an extremely small aperture and thus requires a high sensitivity detector or small magnification to produce sufficient flux at the detector plane. In 1952 Wolter described grazing incidence axisymmetric reflection microscopes and telescopes. X-ray telescopes have been fabricated and fielded in x-ray astronomy programmes and have provided significant information on spatial irradiance distribution of astronomical x-ray emitting sources. The configuration of one of the Wolter x-ray mirror pairs for microscope application is illustrated in Figure 10.6. The first reflecting surface is a hyperboloid of revolution and the second one an ellipsoid of revolution. The common foci of the ellipsoid and hyperboloid are the location of the virtual image. The object is placed at the second focus of the hyperboloid and the film image plane is located at the second focus of the ellipsoid. The imaging system provides perfect geometrical imaging for a point source located at the second focus of the hyperboloid to the image focus of the ellipsoid.

In 1977 a critical study of this system was made to determine its depth of field and imaging capability for off-axis points (*i. e.* sources). It was found that it was theoretically possible to produce submicron imaging over a field of view of 0.5 mm. However, a practical system with an object distance of 30 cm requires fabrication tolerances for 1 μm

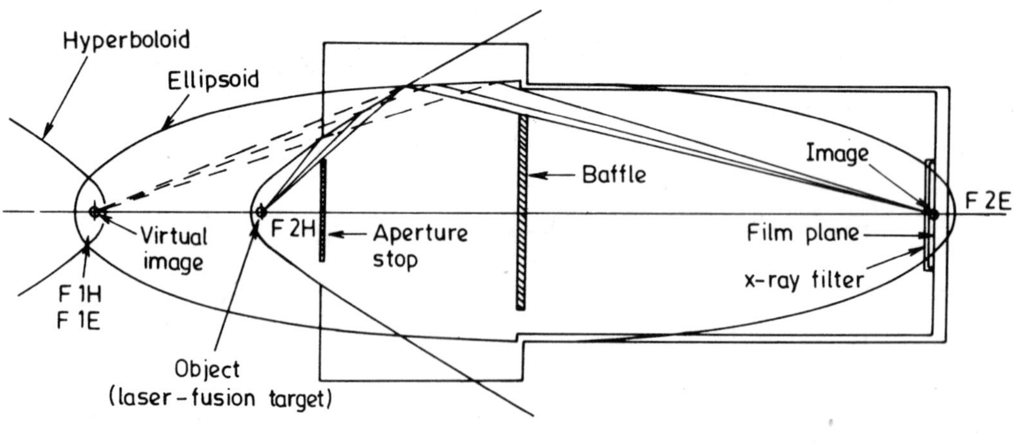

Figure 10.6 The optical configuration of the Wolter Type I x-ray microscope consists of hyperboloidal and ellipsoidal surfaces

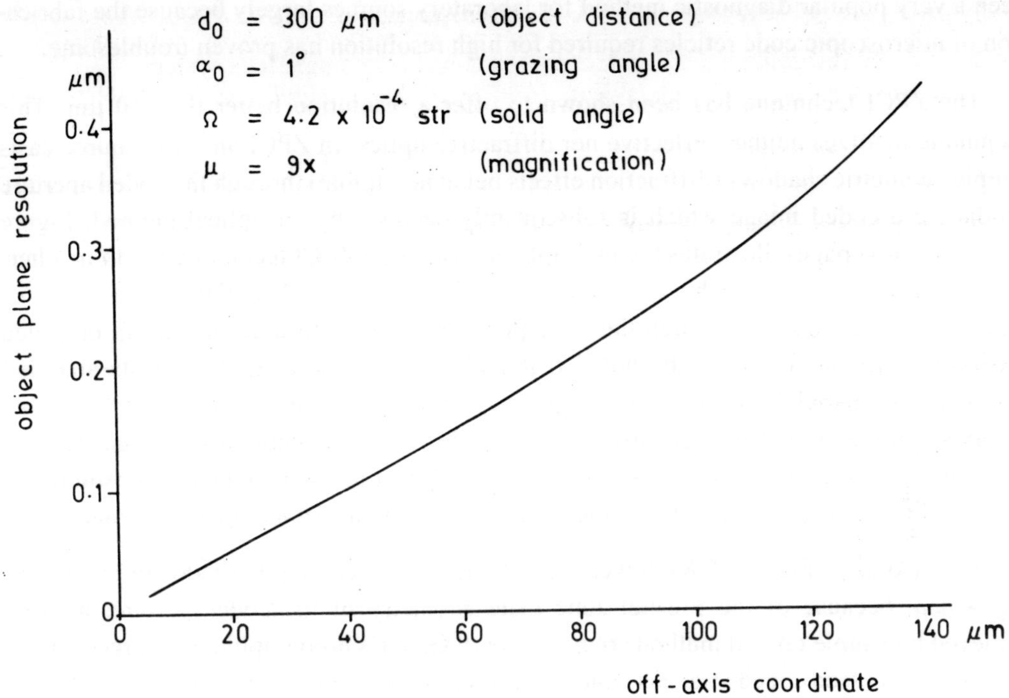

Figure 10.7 Hyperboloid-ellipsoid x-ray microscope resolution

resolution which are beyond the present day state of the art. Surface figure requirements are less than 1000 nm and slope error requirement on the surface is 16.7 nm/cm.

Calculations have been made using a geometrical ray tracing code for an object distance of 30 cm and magnification 9, an x-ray grazing incidence angle of 1° and a subtended solid angle of 4.2×10^{-4} Sr. Computationally, the on-axis object point, which geometrically is perfectly imaged, generated an image point less than 10^{-3} μ m in radius. The results are shown in Figure 10.7.

A mirror has been fabricated at LLNL using a single point diamond turning on a numerically controlled cutbearing lathe. The machine can give surface accuracies of 25 nm; however, roughness is also of the same order, and thus the surface is not an efficient specular reflector of the x-ray wavelength range desired. For 90% specular reflection of 1 KeV x-rays at 1° angle of incidence, the roughness must be less than 1.7 nm.

10.8 Zone plate-coded imaging *(ZPCI)*

Coded imaging techniques have been used in astronomy and nuclear medicine for imaging short wavelength radiations, not easily focussed by using conventional reflection, refraction or diffraction techniques. In spite of its potential coded imaging has not in the past been a very popular diagnostic method for laboratory sources largely because the fabrication of microscopic code reticles required for high resolution has proven troublesome.

The ZPCI technique has been shown to offer a resolution better than 10 μm. This technique involves neither reflective nor diffractive optics. In ZPCI an x-ray source casts simple geometric shadows (diffraction effects being negligible) through the coded aperture producing a coded image which is subsequently decoded by an optical method. Figure 10.8a (*see* next page) illustrates the principle of a two-step ZPCI technique for an incoherent laboratory source distribution of points of short wavelength radiation. Each source casts its distinct shadow through the zone plate aperture on to a shadowgraph or coded recording medium. It should be noted that each shadow characterizes, by its size and position, the spatial location of its associated source point: an off-axis point casts an off-axis shadow, a distant point casts a small shadow, a close point casts a large shadow. Thus the spatial distribution of the source is recorded in the distribution — position and size — of the zone plate shadows in the coded image (Figure 10.8a, upper diagram).

The next step (Figure 10.8a, lower diagram) in this processing is shadowgraph decoding which, because of the Fresnel zone plate geometry of the code aperture, may be achieved by simple optical methods reminiscent of Gabor's holographic image reconstruction. The processed coded image is coherently illuminated using an optical laser. Each zone plate shadow focusses the incident radiation to a diffraction-limited spot, the image of its associated source point, thereby reconstructing the original source distribution point by point. This reconstruction stage uses the diffractive properties of the zone plate to form the image. The reconstruction must be carried out with radiation of much longer wavelength than that used in encoding so that the image will give appreciable diffraction effects. Diffraction theory shows that the light incident on a zone plate is diffracted into orders having real and virtual foci. The focal length for the first order is $f = r_1^2 / \lambda$ where r_1 is the radius of the first zone of the zone plate and λ the wavelength of the reconstructing radiation.

An actual system for reconstruction is shown in Figure 10.8b. It uses spatially filtered HeNe radiation to illuminate the coded image. The reduction in the signal-to-noise ratio due to diffracted light is overcome by using a lens in front of the zone plate. The undiffracted light is brought to a focus in the Fourier plane of the lens and removed with a small stop.

The diffraction pattern downstream from the shadowgraph (*i. e.* coded image), when observed in the proper reconstruction plane, is a magnified image of the source distribution. A proper reconstruction plane for the first order real image is found at a distance *d* downstream from the shadowgraph. For a parallel beam illumination of the coded image we have

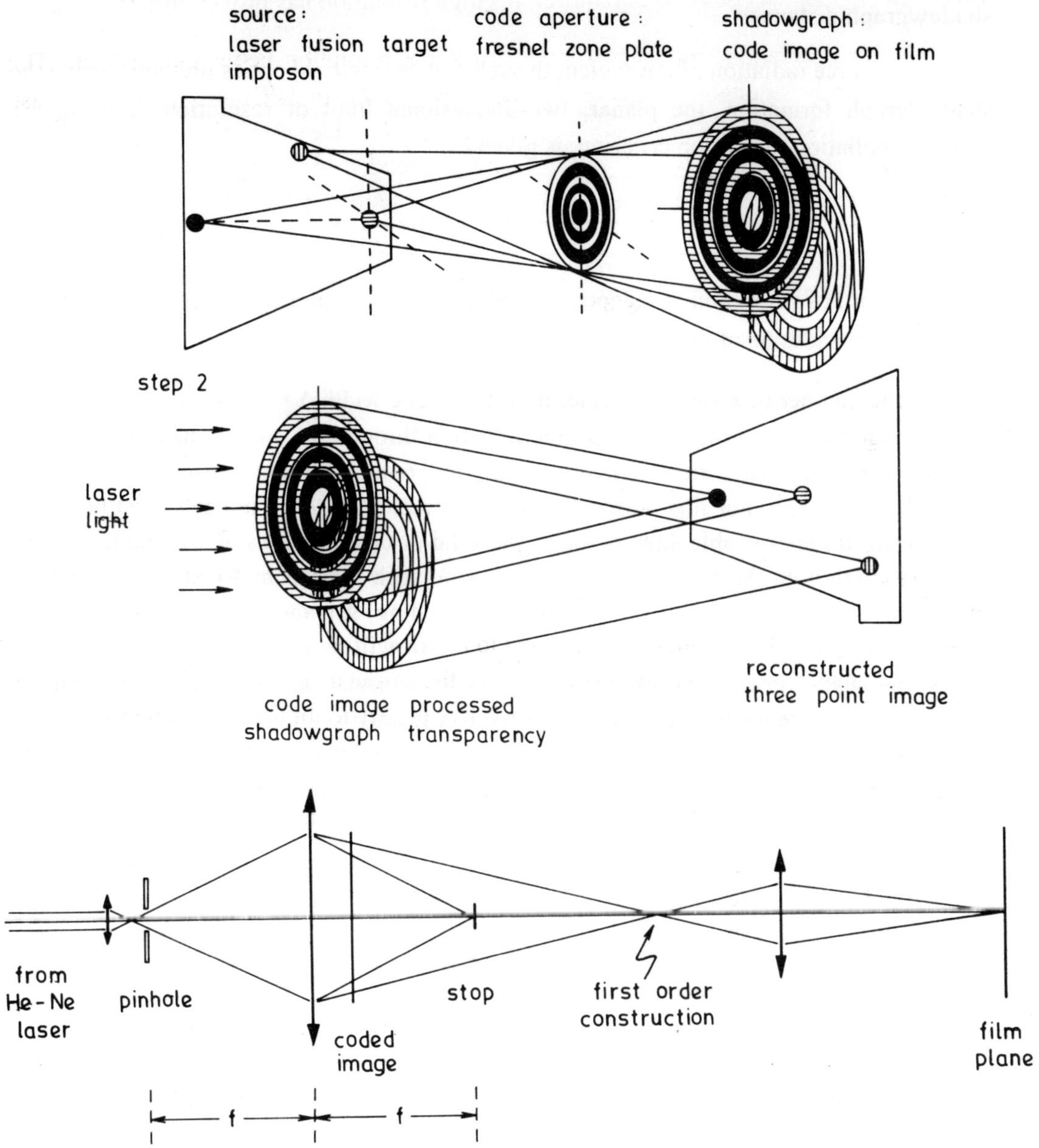

Figure 10.8 (a) Zone-plate coded imaging principle, (b) Reconstructon of image from coded image

$$d = \left(\frac{S_1 + S_2}{S_1} \right)^2 \frac{r_1^2}{\lambda} = (1 + \mu)^2 \frac{r_1^2}{\lambda} \qquad 10.14$$

where $\mu \equiv S_2/S_1$ is the magnification, the ratio of source-to-zone-plate and zone-plate-to shadowgraph distances.

For a source radiation of a wavelength such that geometrical optics prevails during the shadowgraph formation, the planar, two-dimensional limit of resolution δ, using the coherent radiation resolution criterion, is given by

$$\delta = 1.64 \frac{S_1 + S_2}{S_2} \Delta r = 1.64 \left(1 + \frac{1}{\mu}\right) \Delta r \qquad 10.15$$

Thus the limit of resolution is independent of the reconstruction wavelength and is determined by the width Δr of the narrowest zone of the zone plate.

A large number of zones and a finer minimum zone width Δr will yield zone plates of higher resolution. Transmission of source radiation through the solid zones of the coded aperture during reconstruction or encoding stage will reduce the signal-to-noise ratio of the reconstructed image. For this reason the source wavelength to be imaged must be sufficiently long for appreciable attenuation in the solid or opaque zones of the aperture. This criterion along with the short wavelength criterion such that geometrical optics prevails during the formation of the shadowgraph (*i. e.* encoding) sets the short wavelength limit. Thus the precise value of this short wavelength limit will be determined by geometrical optics, the shadowgraph contrast required and the thickness and material of the opaque zones. A zone plate made of gold can improve its capacity to higher x-ray energies which will give better image contrast. In case the source radiation wavelengths are too long to fulfil the geometrical optics condition the resolution limit is given by

$$\delta = 3.3 \left[\frac{S_1 (S_1 + S_2)}{S_2} \lambda_\mu\right]^{\frac{1}{2}} \qquad 10.16$$

where λ_μ is the longest wavelength to be imaged. The long wavelength limit can then be written as

$$\lambda_\mu \approx \frac{(\Delta r)^2}{4} \left(\frac{S_1 + S_2}{S_1 S_2}\right) \qquad 10.17$$

A zone plate-coded aperture is a large aperture device and hence has a large radiation collection efficiency. This is a very important advantage of ZPCI over other imaging methods. The radiation collection solid angle for a zone plate is four to six orders of magnitude greater than a pinhole camera for a similar resolution. ZPCI is capable of three-dimensional imaging. The coded image records the three-dimensional spatial information contained in the source distribution. This information is gathered by

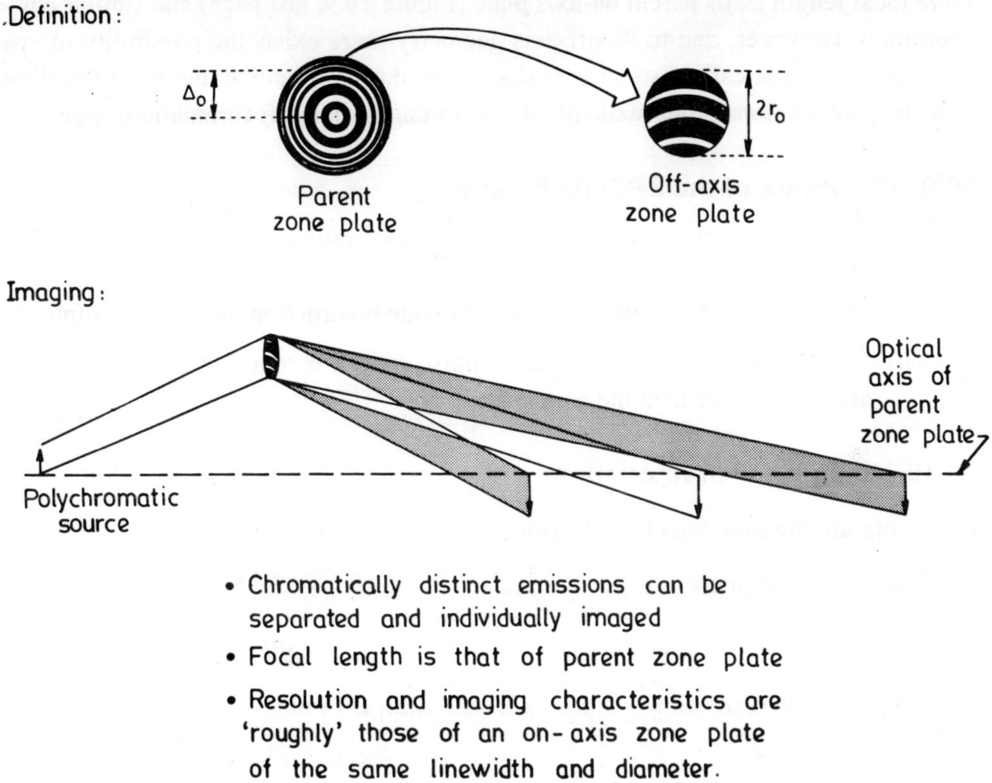

Figure 10.9 Off-axis Fresnel zone plate: Imaging properties

viewing the reconstructed image in a separate reconstruction plane. This is called *tomography*, *i. e.* "writing by sectioning". A quantitative measure of the tomographic capability of ZPCI is given by its tomographic resolution limit:

$$\Delta = \frac{0.75}{N}\frac{S_1}{S_2}(S_1 + S_2) \qquad 10.18$$

where N is the total number of zones.

Since no reflection, refraction or diffraction processes are involved in the first stage of shadowcasting in ZPCI it can in principle be used for any radiation satisfying ray optics. In fact the first demonstration of the capability of ZPCI was charged particle imaging from an electron emitting source.

Off-axis microzone plates are also used sometimes. These can be used as imaging spectrometers for polychromatic x-ray sources. An off-axis zone plate is simply a circular, off-axis segment of a circularly symmetric or *on-axis* zone plate. The off-axis plate has the same focal length as its parent on-axis plate (Figure 10.9, last page) and similar chromatic aberration. However, due to its off-axis geometry there exists the possibility of spatially separating chromatically distinct emissions from the same source, thereby providing multiple, two-dimensional, chromatically distinct images of a polychromatic source.

10.9 Limitations of the ZPCI technique

(*i*) Since no ray tracing is involved in shadowgraph recording the source emission should be diffuse. Each point in the source must illuminate the entire coded aperture in order properly to have its code recorded in the shadowgraph.

(*ii*) In order to have a good signal-to-noise ratio it is required that the active source area be smaller than the zone plate area.

10.10 Imaging characteristics

Following are the imaging characteristics of a typical ZPCI setup:

Effective solid angle .... $10^{-2} - 10^{-1}$ Sr

Planar resolution .... 5 – 10 μm

Tomographic resolution .... ≥ 10 μm

Field of view .... order of a few cms

Spectral range .... 3 – 50 KeV

Typical object distance .... 1 – 2 cm

10.11 X-ray Fresnel lenses

We can also use Fresnel zone plates in the x-ray diffracting mode in which case they become x-ray lenses. In this mode 10% of the incident radiation is diffracted into the first order real focus. The focal length is given by $f = r_1^2 / \lambda$. The resolution of an FZP is given by

$$\delta = 1.22\ (1 + 1/\mu)\ \Delta r \qquad 10.19$$

μ being the magnification and Δr the width of the outermost zone. For a geometric zone plate the outermost (*i. e.* smallest) zone width may be simply expressed as

$$\Delta r = (r_N) / (2N) \qquad 10.20$$

where N is the total number of zones and r_N the overall ZP radius; thus the f/number (*i. e.* focal length/diameter) of the zone plate depends only on its smallest zone width and the x-ray wavelength used: $f^{\#} = (\Delta r) / \lambda$. Chromatic aberration may also be used to advantage as discussed earlier in the case of the off-axis segment of an FZP.

10.12 X-ray microscopy using FZP

An example of this type of utilization is shown in Figure 10.10. In this case an FZP with a minimum Δr of 320 nm was used to image another FZP with a minimum Δr of 1 μm backlit by x-rays produced by a proton beam incident on an Al target. Microdensitometry of the exposed image shows that it was possible clearly to resolve the minimum zone width of the x-ray-backlit ZP, thus demonstrating a resolution of better than 1 μm. This offers great flexibility and possibilities for x-ray optics in the future. However, present versions of the zone plates for this type of application are not very stable.

The demonstration of diffraction-limited resolution of FZP lensing elements encourages the design of the next generation FZPs, which will bring us into the regime of x-ray microscopy and microanalysis on a sub-100 nm scale. A number of criteria stand out in the design of such an optical component, *e. g.* distortion in the pattern generation and the problem of mechanical stability of the free standing structure. To keep the size small one

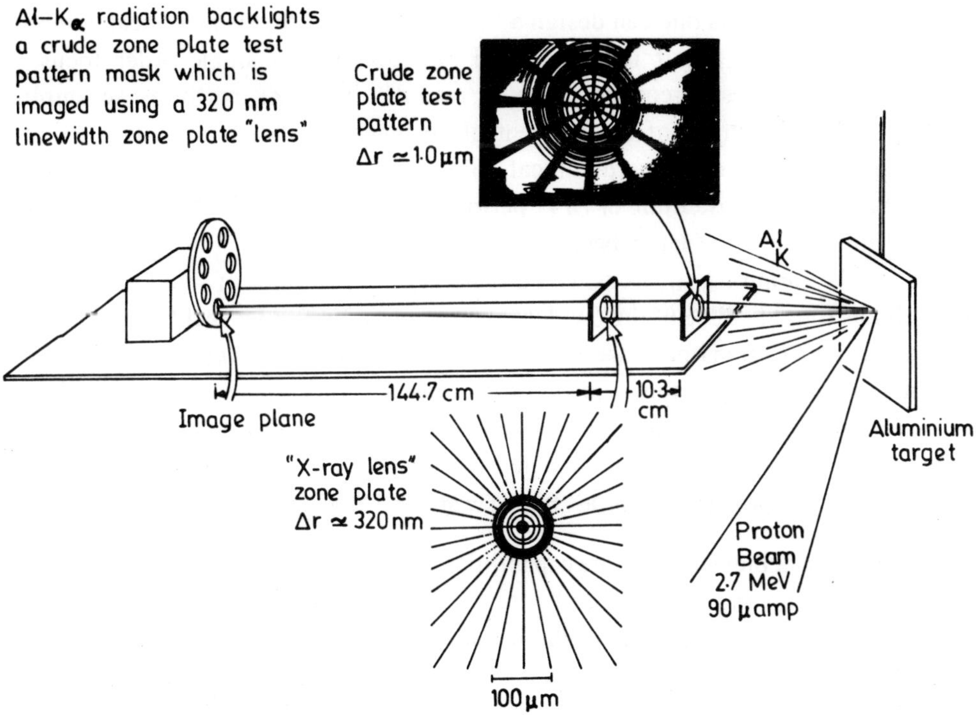

Figure 10.10 Using a 320 nm-linewidth Fresnel zone plate as a 14 × microscope the 1.0 μm resolution has been demonstrated with Al-K_α radiation

may choose N ≈ 100 and for good resolution r ≈ 500 nm. Such a lensing element would produce the sub-100 nm probe spot size of interest in x-ray microanalysis while taking a significant step towards the goal of about 20 nm resolution, at which the microscopy of live microbiological samples becomes most interesting. The parameters Δr = 50 nm and N = 100 set the FZP diameter at 20 μm. The zone plate thickness t required as well as its focal length and f/number are established by the choice of the x-ray wavelength. At λ = 5 nm the focal length is 200 μm, f/number 10 and a material thickness t = 100 nm of gold, for example, is sufficient to absorb more than 90% of the radiation incident on the solid zones.

A significant feature of the discussion above is that in the end the matter of feasibility reduces to lithography techniques. Perhaps the most interesting challenge ever confronted in x-ray microscopy reduces quite simply to microstructure fabrication. Can a microscopic pattern of concentric circles be written with a precision of 20-50 nm? This will probably be answered in the coming few years.

10.13 Immediate Future

Significant increase in x-ray lens efficiency and background reduction (due to other orders) can be achieved by employing phase modulation effects in diffractive optical components. Employing such concepts one can design a "true x-ray lens" that is an optical component which focusses nearly all of its transmitted radiation into a single order focus. The evolution of such a design concept is based on FZP. The FZP performs an amplitude modulation (periodic in r^2) of the incident plane wave, focussing roughly 10% of the incident energy into the first order focal spot. A somewhat more efficient focussing element, the *Fresnel Phase Plate* or *FPP*, performs phase and amplitude modulation of the incident plane wave – the latter because of the finite x-ray absorption. The modulation function has the form of a square wave periodic in r^2. In the limit of small absorption the FPP can focus roughly 40% of the incident energy into a first order real focus. The blazed Fresnel plate *(BFPP)* transforms the incident plane wave into a converging spherical wave having an amplitude modulation (-due to finite x-ray absorption) periodic in r^2. In the case of small absorption practically all of the unabsorbed energy is directed into the first order real focus. The phase modulation elements can be made of aluminium. An aluminium thickness of 1.6 μm produces a phase shift of π at λ = 1 nm.

The key to the achievement of the capabilities mentioned above lies in the achievement of the technology for the fabrication of periodic microstructures with precision thickness and profile (for the Brazed Fresnel Phase Plate) from materials chosen for their x-ray refractive properties. BFPP still appears to be a difficult device to make. However, x-ray FPP could indeed be fabricated from a variety of materials (Al, Cu, Si, polymers, *etc*) with near term technological efforts under way.

Chapter 11

PROGRESS IN PATTERN RECOGNITION AND IMAGE PROCESSING

B L DEEKSHĀTULU and R KRISHNAN

11.1 Introduction

Pattern recognition is now an area of intense theoretical and applied research. The nearly quarter century of activity in this area has produced a few spectacular results but has also had a sobering effect in making us realize the inherent complexity of this activity. In fact the progress in certain areas of pattern recognition is minuscule when viewed against developments in computer hardware or even in systems software.

Statistical and structural pattern recognition approaches have received much attention. The extension of formal language ideas to two dimensional pictures prompted the development of web grammars and picture description languages; however, these tools have not proved very successful and structural or syntactic approach has not lived up to expectations.

Statistical approaches have had relatively greater success. Low level image processing especially has succeeded in developing several powerful operators. Higher level image processing has dealt with segmentation and image understanding.

Today there is a growing realization that perhaps the superiority of the human brain in pattern recognition is due not to the presence of some exotic pattern recognition operator but probably due to its ability to decide which tools to use and in what combination. This has resulted in attempts at using heuristic methods under the title of *knowledge-based systems*.

Regarding application there has been moderate success in biomedical image processing and remote sensing. Industrial applications like automated visual inspection have had limited success.

Computer vision has emerged as an independent discipline and hence is not covered in this article. Five different areas have been identified and a brief report of some recent activities is presented. These are: (*i*) Architectures for image processing, (*ii*) Quadtrees, (*iii*) Texture, (*iv*) Segmentation and edge detection, (*v*) Artificial intelligence in pattern recognition.

11.2 Architectures

Special architectures for image processing are increasingly assuming significance. The image processing operations can be grouped into iconic and symbolic. Iconic operations are pixel-based operations whereas symbolic ones are higher level operations. For iconic operations several architectures have been proposed; for symbolic ones a serial computer is still the best choice.

All of the iconic architectures can be said to belong to the general class of SIMD machines. The three most popular ones are (*i*) pyramid, (*ii*) linear array, (*iii*) N-cube.

In a pyramid (Figure 11.1) the apex is a single processor. At each level the processors communicate with other processors at the same level, with a few processors at the next level and with one processor at the next upper level.

Under this basic scheme various variations occur. Tanimoto (1983) and Cantonic *et al* (1985) have suggested that all processors should be bit-serial and that a quadtree paradigm should be used for interconnection. Uhr's (1981) system has a hierarchy of different processors at various levels (Fountain 1986). Scene Analysis is one of the well known areas for applying pyramids. The vision cone is an artefact used to explain the pyramidal

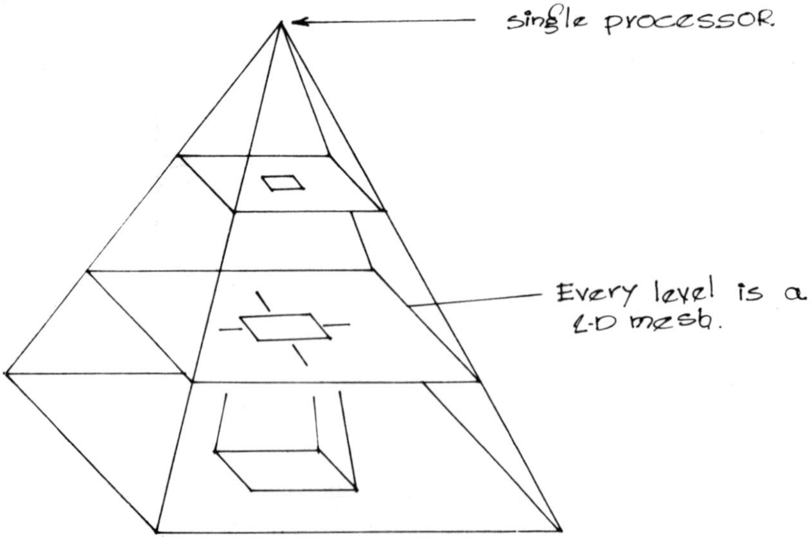

Figure 11.1 Pyramid

processing. At the lowest level of the pyramid basic processing operations are carried out (edge detection, texture identification, *etc*). At higher levels more complex abstractions are

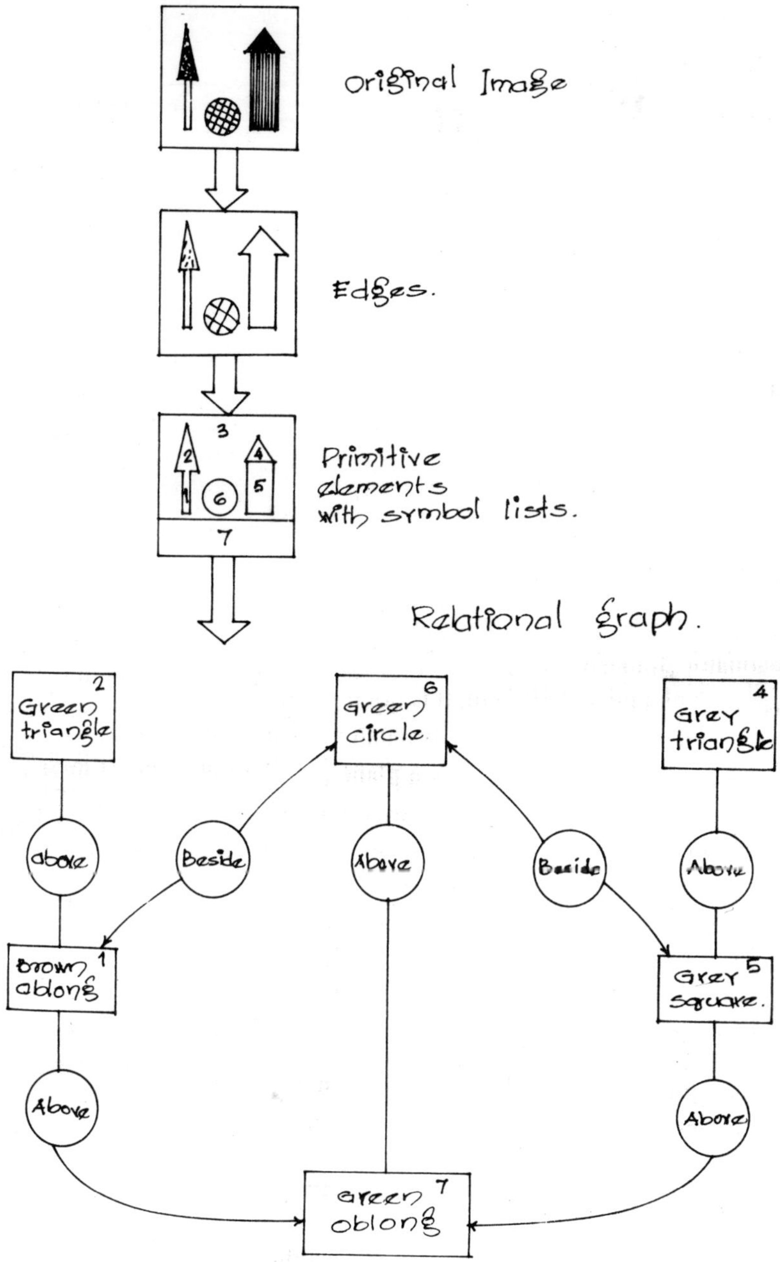

Figure 11.2 Pyramid for analysis

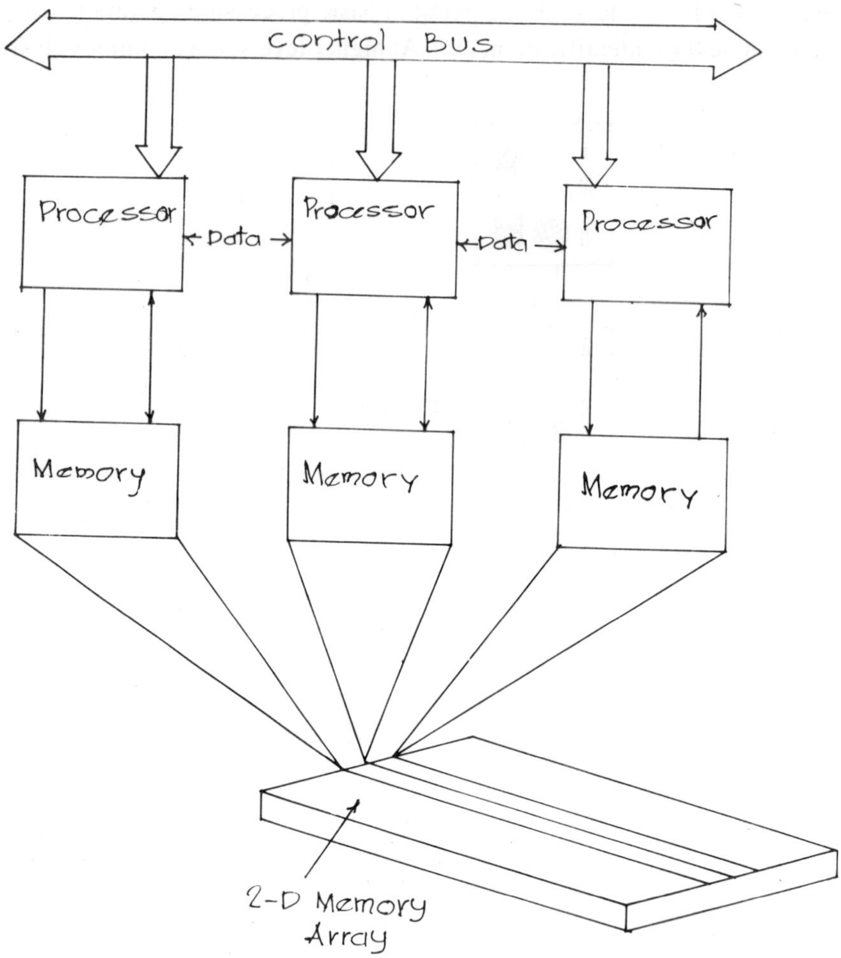

Figure 11.3 Linear array

derived culminating in a relational graph at the apex. Figure 11.2 (*see* p. 161) from Fountain (1986) illustrates this concept.

Figure 11.3 shows the basic features of the linear array. Here the processing elements are arranged linearly. The associated memory is usually two-dimensional implying that 2-D data can be handled. The processing elements are multibit and high bandwidth communication between processors is provided for.

ZMOB was the first such system. It consisted of 280 microprocessors. More recent suggestions are the WARP, a ten element array, a 256-element system based on the CLIP7 chip, PICAP3 and SLAP. These systems differ basically in their scale and the degree of autonomy permitted each processor.

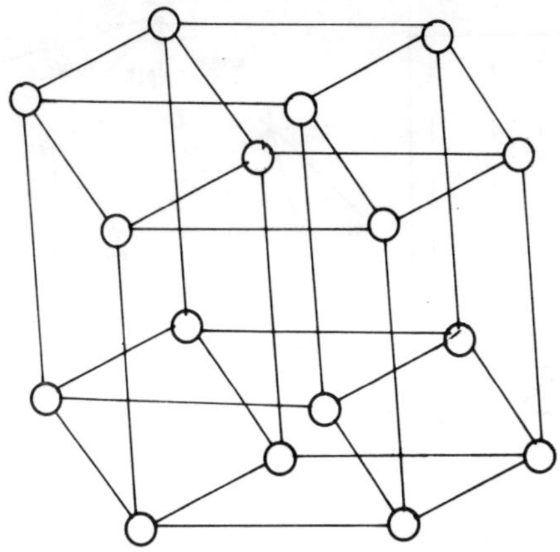

Figure 11.4 Cube

11.3 N-Cubes

In the N-Cube architecture (Figure 11.4) the processors are interconnected in such a way that no pair of processors is separated by more than 2logN links. Even though this type of connectivity is the basis of N-cube architecture the commercially developed systems have slightly different features. INTEL's 1iPSc developed from Caltech's Cosmic cube and the Floating Point system T-series place a powerful microprocessor at each node of the network. The connection machine starts with a large two-dimensional array of bit-serial processors. Each 4 × 4 subarray of these is served by a message-passing device and it is these which sit at the nodes of an N-cube network.

For low level operations like edge detection both pyramid and N-cube perform equally fast while the linear array takes about ten times more time. For data shifting which is a second level operation the N-cube is much superior and the pyramid is the second best.

Hence, considering the three architectures, the pyramid and the N-cube are superior in performance and cost compared to the linear array.

11.4 Quadtrees

The quadtree representation of a binary image has found applications in image processing, computer graphics, robotics and pictorial information retrieval. It is also used as a data compression device. For iconic image processing some of the architectures proposed use quadtrees for interconnection.

Consider Figure 11.5 (*see* next page). The figure on the right shows the quadtree representation of the figure on the left. G stands for grey, B for black and W for white. The

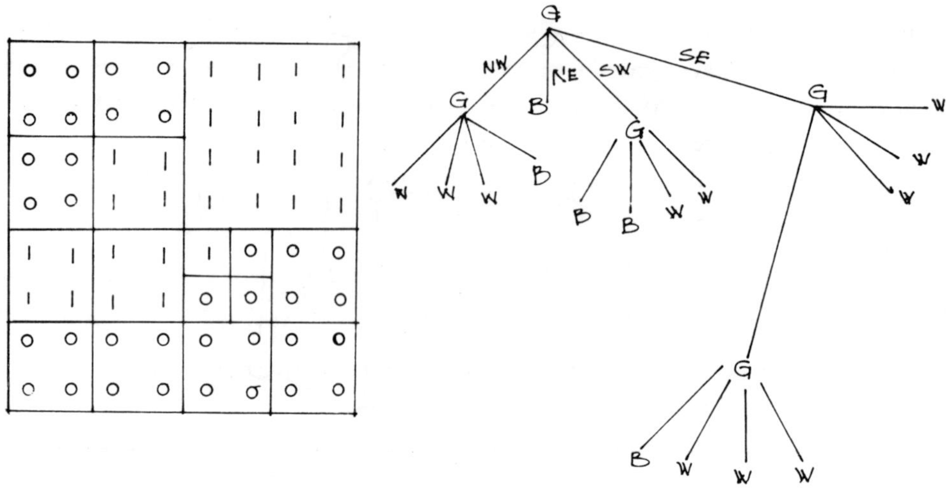

Figure 11.5 Quad tree

four directions are the northwest, northeast, southwest and southeast. By successively dividing the image into quadrants the quadtree is constructed. The original representations of quadtrees were pointed representations.

Recently there has been considerable interest in pointless representations called *linear quadtrees*. One such is called the DF representation in which a pre-ordered traversal of the quadtree which yields a string is generated. For Figure 11.5 the DF expression is

((00011 ((1100 ((1000000

where "(" stands for grey, "0" for white and "1" for black. Another recently proposed representation (Qing Yun Shi 1986)is the CD ("colour depth") representation which facilitates image processing operations and also does not require more space than the DF. Let P = b(i,j), i, j = 1, ..., 2^L be a binary image of $2^L \times 2^L$ pixels. While tracing its quadtree in preorder we successively write down the colour and depth of each leaf node. This process will yield an attributed string (bi,di) ... (bm,dm), which is called the CD representation of P. For example, the CD representation for this image is

(0,2)(0,2)(0,2)(1,2)(1,1)(1,2)(1,2)(0,2)(0,2)(1,3)(0,3)

(0,3)(0,3)(0,2)(0,2)

Another recent algorithm by Shaffer and Samet (1986) builds a linear quadtree from a raster image stored on disk in time proportional to the number of nodes in the output quadtrees. This is an order of magnitude better than existing algorithms. Another representation of quadtree which is useful in geographical information systems is the PM quadtree.

11.5 Texture

Texture is important for both classification and segmentation. Both statistical and non-statistical approaches have been taken. Statistical methods have by far received the most attention. The popularity of features based on second order statistics originate from the well known conjecture by Julesz (1962) that human beings cannot distinguish between textures which do not differ up to second order statistics. This conjecture still seems valid for natural textures despite a few counter examples.

The grey level occurrence matrix has been the most successful representation for statistical texture. Several measures like contrast have been defined using this matrix.

Fourier spectra and run length matrices, spatial domain filter masks or texture field models like autoregression or Markov random field models are the other devices used for texture analysis.

Texture measures have been used for classification also. The co-occurrence matrix is subdivided into squares and the average of each square is taken as a feature and the features together form the feature vector. This feature vector is used to classify the texture using what are known as subspace methods. Identifying texture boundaries in a digital picture is essential for picture segmentation. This is usually a difficult problem. Davis and Tyehon

Figure 11.6 Original textured image

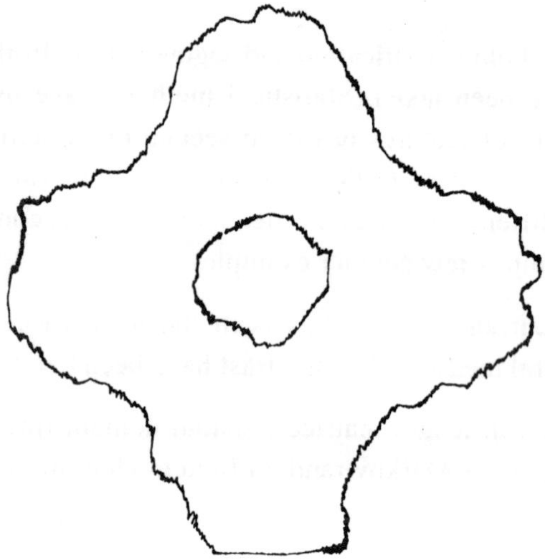

Figure 11.7 Edges obtained

(1986) have approached this problem by defining higher order pictures. They suggest that textural measurements be generated which will themselves form a picture which, then, can be processed in the usual ways. We shall briefly explain this procedure: Let P be the set on n points to be analyzed and D the set of n adjacent differences. Thus

$$P = p(1),\ p(2),\ \cdots,\ p(n),\ and$$

$$D = d(1),\ d(2),\ \cdots,\ d(n-1),\ where\ d(i) = p(i) - p(i+1)$$

One measure determines the frequency characteristics of the picture. It counts the differences which are above a threshold and one of a different sign than the previous difference. The second measure captures the information on contrast in a texture. It is defined as

$$C = \frac{\Sigma\ |d(i)|}{(n-1)}$$

These two measures are used to analyze four vectors about each point in a textured scene, where one vector exists for each orientation. This 8-feature vector at each point is used to represent a second order picture. By carrying out edge detection operation in this picture textures are distinguished (Figures 11.6-7; *see* former on last page).

Kasvand (1986) has drawn attention to the textured lines (dotted, dashed, *etc*) used in engineering drawings but are rare in nature. Figure 11.8 (*see* next page) shows textured

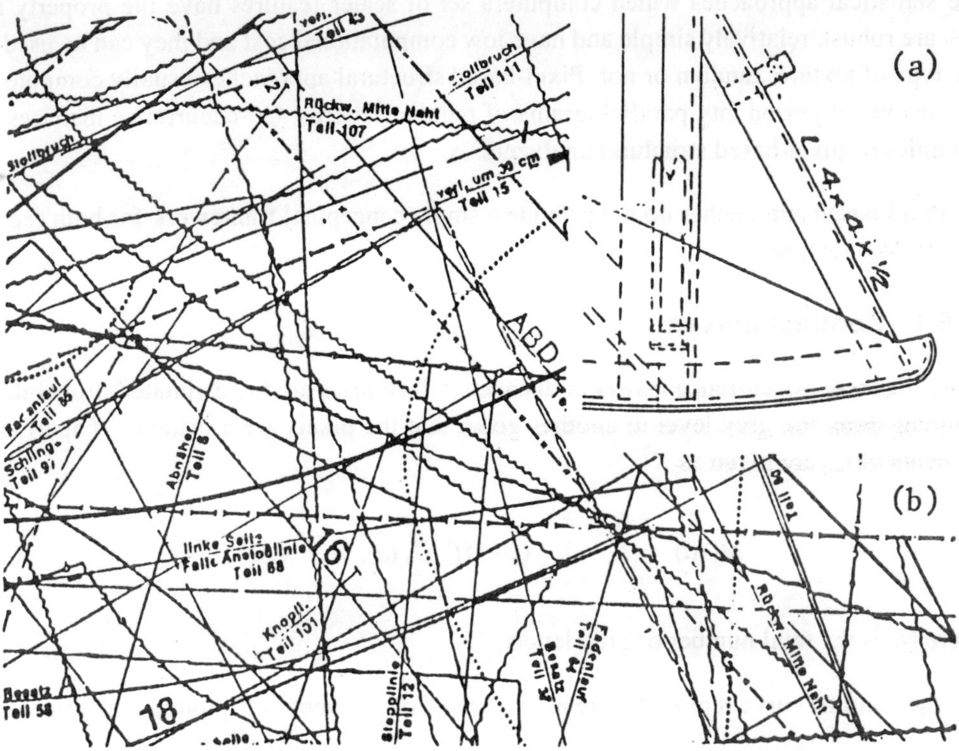

Figure 11.8 Examples of textured lines. (a) A portion of an engineering drawing. The lines are usually black and of approximately equal thickness. (b) A portion of a dress pattern. The lines on the original are in colour.

lines. An elaborate set of steps is proposed to identify line segments, junctions, *etc*, and finally cluster the textures.

Colour textures have received relatively less attention. One method is to combine the Red, Green and Blue images linearly with weights determined from psychovisual experiments and then deal with a grey picture. Another attempt has been to increase the number of grey levels by combining the three colours (*i. e.* to have 256^3 grey levels and cluster the data in R-G-B space).

Abstract statistical approaches to texture-compute a set of scaler features describing the distribution of irradiances or local features but ignoring their spatial dependence. Structural approaches concentrate on the spatial interaction of elementary regions, local features or irradiances. Most of them assume that texture pattern being analyzed is highly regular and fail to cope with the gradual transition from regularity to randomness. The contrast curve (the first moment of the co-occurrence matrix plotted as a function of the intersample distance) as a texture descriptor, has been suggested by Chetverikov (1984).

The statistical approaches which compute a set of scaler features have the property that they are robust, relatively simple and have low computational cost and they can be used for any type of texture, random or not. Pixel-based structural approaches usually compute the dimensions of periodicity parallelograms of regular patterns. Co-occurrence matrices are examples of pixel-based structural analyzers.

Pixel-based approaches do not provide a single conceptual framework for both regular and random textures.

11.5.1 Contrast curves

Let C_{ij} be the co-occurrence matrix at distance *d*. Its entries are the estimated probabilities of going from one grey level to another given that the pixels are a distance *d* apart. The moments of C_{ij} are given as

$$M_n(d) = \sum_{i-1}^{N_g} \sum_{j=1}^{N_g} (i - j)^n \, C_{ij}(d), \; n = 1, 2$$

where N_g is the total number of grey levels.

The contrast curve represents the normalized first moment of C plotted a function of d:

$$CON(d) = \frac{M_1(d)}{i_{\max} - i_{\min} + 1}$$

where i_{max} and i_{min} are the maximum and minimum grey levels in the image. Periodic textures have a periodic contrast curve which reduces in amplitude as the distance *d* increases. The curve of a pattern possessing a very weak regularity has a single maximum. A random pattern is described by a monotone contrast curve.

Textural patterns called *zebra patterns* can be used to describe textural patterns. Consider pulses of unit height but random length. These represent zebra patterns. Figure 11.9 (*see* next page) shows the contrast curves for three different patterns: netting, coffee beans and calf leather.

Edge detection and orientation are tools which are useful for texture discrimination. The high frequency points can be classified into (*i*) directional edge points, (*ii*) isotropic edge points, (*iii*) non-edge points. A feature vector $X = (N_1, N_2, ..., N_g)T$ with $N_i = N^2$ can be defined for a $N \times N$ region in the picture where $N_1, N_2, ...$ represent the number of points which belong to the different directional classes. Using training textures, unknown textures can be classified.

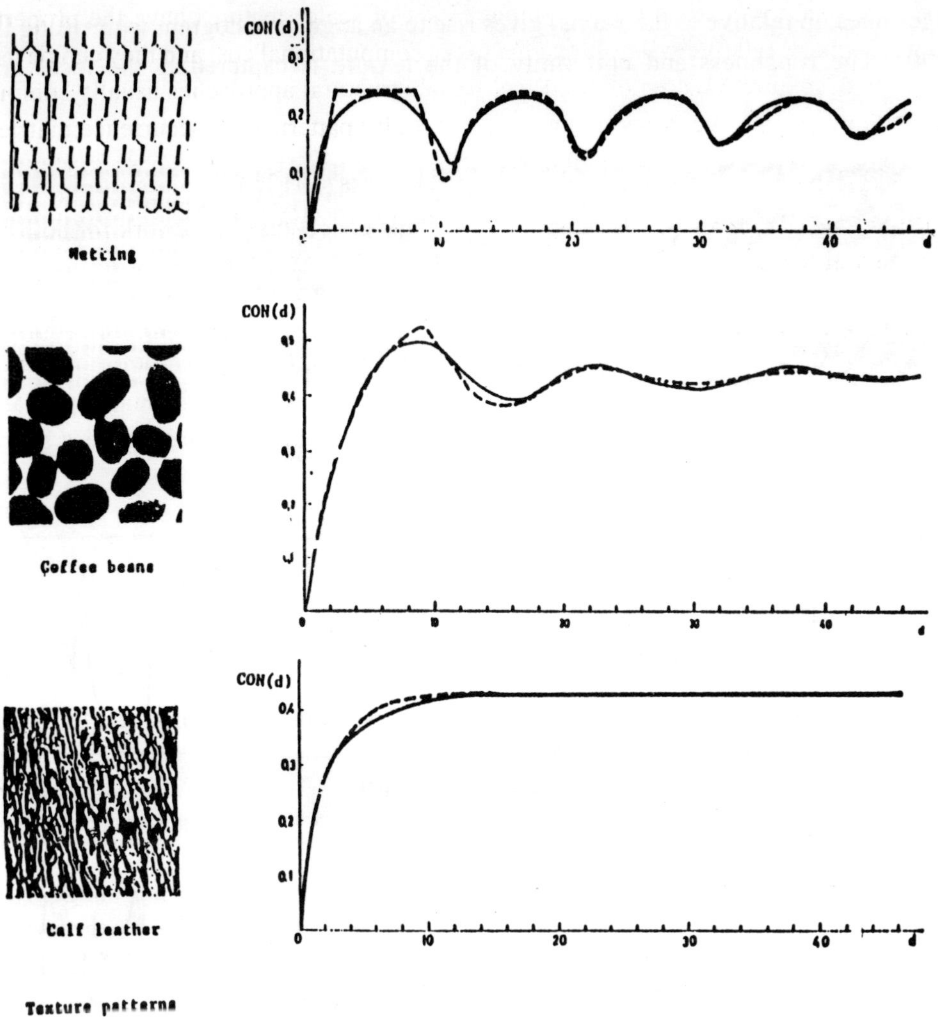

Figure 11.9 The best-fitting zebra curves, dotted, superimposed on the real curves

11.6 Texture synthesis

Texture synthesis is essential for flight simulators or for cinematic effects. To construct an image from compressed data, texture synthesis is essential. By representing a texture by a stochastic process textures can be synthesized to resemble natural textures.

Local histograms of image grey levels capture the tonal variations in the image but do not adequately represent textural properties. To provide additional textural information local histograms over other image measures such as the gradient magnitude and gradient angle may be needed. For example, a texture with a strong *directional* component (*i. e.* with edges lines up relative to the x-axis) gives rise to an angular histogram with strong lines at $\pm$ 90°. The roughness and uniformity of the texture is captured in the local gradient

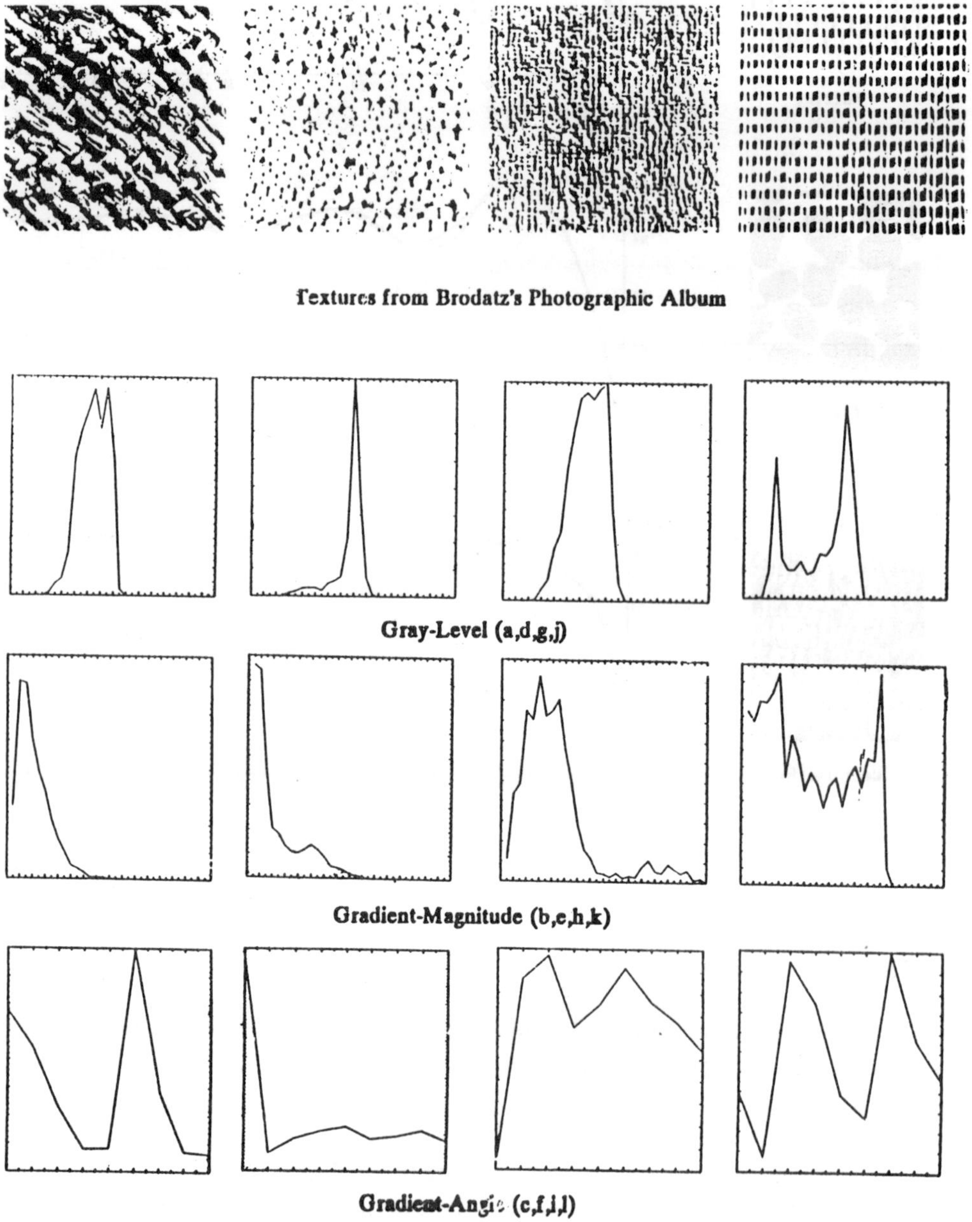

Figure 11.10 Reference histograms

magnitude histogram. The position of the major mode will be indicative of the average roughness. The uniformity (variation in the roughness) will be proportional to the spread in the histogram as shown in Figure 11.10 on the last page (Carlotto 1984).

11.7 Segmentation and edge detection

Image segmentation tries to partition the image into regions which are uniform in some sense, and to find the relation between these regions. Edge detection was a tool used initially to achieve this aim. Edge detectors locate the boundary between regions using the gradient operators. However, these edges rarely look like clear thin boundaries; they often resemble stripes; this is because the edge location operators usually take local statistical variations which are independent of the *global* boundaries. Some work has been done to find global edge detectors but with limited success only.

Thresholding has been the other tool used in segmentation; here the grey values are compared with a threshold and a binary picture is generated. Thresholding has been quite successful because closed digital curves are produced by this procedure. These thresholds are found using the histogram. The histogram minima correspond to the gradient maxima and can be used as thresholds. These histogram methods together with edge detection procedures have proved quite successful. In one method only the histogram of the edge points is taken for thresholding.

These methods fail when the picture is shaded or when the intraregional variation is great. Palvidis (1977) suggested what he calls a "Region Uniformity Predicate". This predicate is true, for example, if any two pixels in a region differ by less than certain grey value *etc*. The socalled split-and-merge technique can be used for segmentation using the uniformity predicate. To start with, the picture is divided into small regions arbitrarily and the uniformity predicate is evaluated inside and between unions of these regions. If any region violates this predicate it is split. If a union satisfies this then the regions are merged. The procedure works well but is time consuming.

Image segmentation using maximum *à posteriori* estimation and Markov random field image models has received much attention recently. Markov random mesh models have also appeared. Let $V_{M,N} = (i, j)$; 1iM, 1jN denote the finite M × N integer lattice. Associated with every pixel (m, n) are a label m, n and a feature vector $X_{m,n}$. Hence $V^{\lambda}_{M,N}$ and $V^{x}_{M,N}$ denote M × N arrays and feature vectors. $V_{M,N}$ is regarded as a realization of a homogeneous, third order Markov mesh random field. For third order Markov mesh random field we have

$$P\left(\lambda_{m,n} \mid \lambda_{ij} : (i,j)\ \varepsilon\ V_{a-1,N}\ U\ V_{M,b-1}\right)$$
$$= P\left(\lambda_{m,n} \mid \lambda_{m-1,n} ; \lambda_{m-1,n-1} ; \lambda_{m,n-1}\right)$$

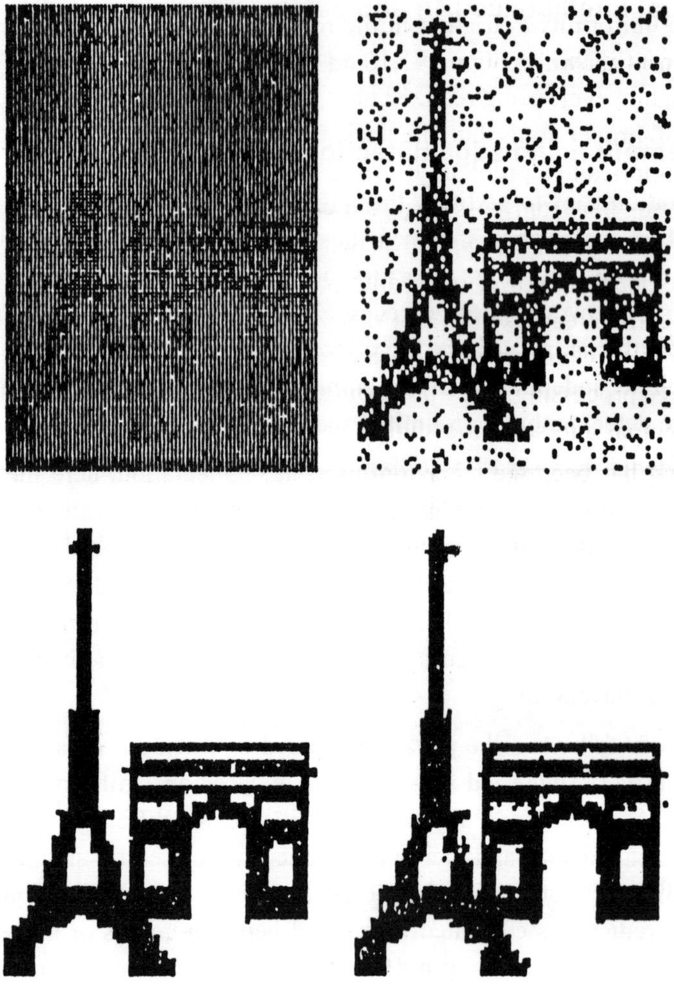

Figure 11.11 *Upper left:* noisy input image (SNR = 1). *Proceeding clockwise,* non-contextual, filtered and smoothed restorations

and similarly for the array.

$$P\ (\lambda_{m,n}\ |\ \lambda_{i,j}\ :\ (i,j)\ \varepsilon\ V_{M,N}\ -\ \{(m,n)\}\ )\ =\ P\,\lambda_{m,n}\begin{vmatrix}\lambda_{m-1,n-1} & \lambda_{m-1,n} & \lambda_{m-1,n+1}\\ \lambda_{m,n-1} & 0 & \lambda_{m,n+1}\\ \lambda_{m+1,n-1} & \lambda_{m+1,n} & \lambda_{m+1,n+1}\end{vmatrix}$$

The problem to be tackled is the extraction of labels, knowing the underlying Markov model and noise model, from a noisy image. Using the known contextual properties this is accomplished; Figure 11.11 shows the results.

11.8 Artificial Intelligence (*AI*) and pattern recognition

Even quite recently the mention of *AI* and pattern recognition together raised a few eyebrows. It is, however, being realized now that today's problem in pattern recognition is not the lack of proper tools but the lack of knowledge about which tools should be combined and applied to a problem. It is believed that the superior human pattern recognition ability comes not from possessing an esoteric single tool but the ability quickly to detect which tools should be applied in what order. Hence attempts are being made to incorporate the heuristic power of AI into pattern recognition. A few systems have been designed in specific areas like the SIGMA system of Matsuyama and Hwang (1985) and Matsuyama (1986). Chassery and Garbay (1986) have given a framework for expert systems in image processing. They point out that the traditional division of knowledge into physical, perceptual and semantic should be changed to descriptive and procedural. Descriptive knowledge is a set of assertions characterizing the morphology of the various entities involved at the different image organization levels. It also includes a description of the spatial relationships characterizing the topography of the image at each level.

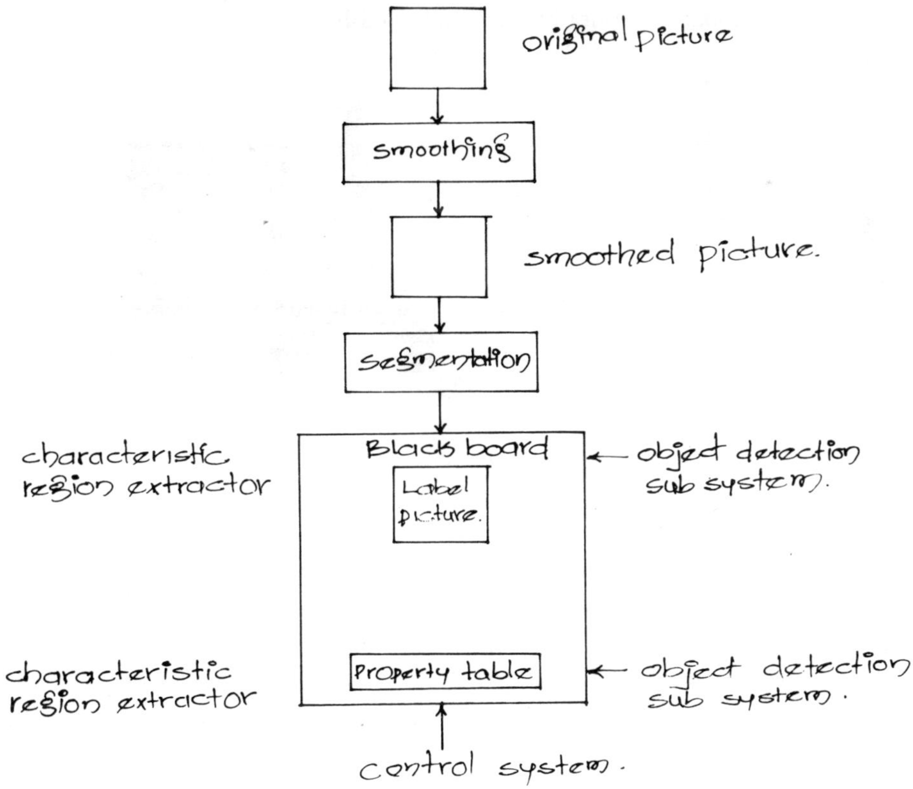

Figure 11.12 Blackboard model

Procedural knowledge indicates the way by which "is-a" relationship and part of a relationship can be obtained.

An expert system approach is useful because of its modularity and heuristic power. Modularity provides easy handling and updating of large data set while heuristic power allows solving the combinatorial explosion of facts by developing adequate reasoning strategies.

Both production-based and frame-based approaches have been thought of.

The problem of combining the LISP/PROLOG-based AI software with the large volume of other language-based image processing software deserves attention.

The actual use of AI in pattern recognition is yet to begin. One area where some success has been achieved is in speech processing. An AI-based speech analysis system has been developed at the University of Turin in Italy. Some other applications have, for example, been in submarine detection seismic signal analysis.

Aerial image understanding is one area where some attempts have been made to incorporate AI techniques. The methodology used is to combine different object detection modules. The "blackboard" model has been adopted by some image understanding systems. Figure 11.12 (*see* last page) shows a system based on the blackboard model.

SIGMA and ACRONYM are systems developed in Tohuku University in Japan (Matsuyama 1986).

11.9 Remote Sensing

Remote sensing is an inherently interdisciplinary field in both its application and its technology base. Hence it has provided motivation for research in diverse areas including pattern recognition (*PR*) and image processing (*IP*). Conversely, developments in PR and IP have generated more uses in remote sensing. It is also generally conceded that, though much progress has been made, the full potential of the space viewpoint for gathering information about the earth and its natural and manmade processes has not yet been thought of, let alone realized. Digital image processing (*DIP*) has been a primary reason for the emergence of remote sensing as an effective means of natural resource assessment. Increasing complexity and decreasing cost of digital hardware enables fast manipulation of data for highest information extraction that would be quite difficult if not impossible with classical photographic interpretation. DIP provides a broad range of capabilities to process and utilize remotely sensed data. DIP can be used to rectify and enhance the data for more effective interpretation. Moreover, it offers a means of information extraction which so very difficult or impossible by photographic or other methods. This is especially true with the development of higher dimensional imaging instruments and integration of data from different sources.

Remote sensing has used pattern recognition in a limited way. Many of the low level IP operations as well as classification procedures have been used. Recently there has been interest in applying AI to remote sensing. One of the first attempts at using AI in PR applied to remote sensing has been done at the Canadian Centre for remote sensing. They have developed modular expert systems for classification, registration, *etc* and image processing and/or classification operations are guided by these systems. Another recent application is in the detection of water bodies (*i. e.* upper reaches of rivers) using AI and Landsat data. The need for handling large volumes of data and for carrying out pattern recognition operations on them has spurred the development of special hardware processors and efficient displays.

References

Cantonic V, Ferretti M, Levialdi S and Maloberti F, 1985 "A Pyramid project Integrated Technology" in *Integrated Technology for Parallel Image Processing* Ed S Levialdi (London: Academic Press) p 121

Carlotto M J, 1984 "Texture Classification based on Hypothesis Testing Approaches" in Proc 7th Int Conf on Pattern Recognition, Montreal, p 93

Cherverikov D, 1984 "Measuring the Degree of Texture Regularity" in Proc 8th Int Conf on Pattern Recognition, Montreal p 80

Chassery J-M and Garbay C, 1986 "Expert Systems, Image Processing and Image Interpretation" in Proc 8th Int Conf on Pattern Recognition, Paris, p 175

Davis W A and Tyehon G G, 1986 "Texture Boundaries in Digital Images" in Proc 8th Int Conf on Pattern Recognition, p 402

Fountain T J, 1986 "Array Architecture for Iconic and Symbolic Processing" in Proc 8th Int Conf on Pattern recognition, Paris, p 24

Julesz B, 1962 "Visual Pattern Discrimination", IRE Trans on Information Theory IT-8, p 84

Kasvand T, 1986 "Linar Textures in Line Drawing" in Proc 8th Int Conf on Pattern Recognition, Paris, p 398

Matsuyama T, 1986 "Present and Future Role of AI in Remote Sensing", Workshop on Analytical Methods in Remote Sensing for Geographic Information Systems, Paris

Matsuyama T and Hwang V, 1985 "SIGMA, a Framework for Image Understanding - Integration of Bottom-up and Top-down Analysis" Proc 8th IJCAI, p 908

Palvidis T, 1977 *Structural Pattern Recognition* (New York: Springer-Verlag)

Qing-Yun Shi, 1986 "Image Processing Operation using CD Representation" in Proc 8th Int Conf on Pattern Recognition, Paris, p 320

Shaffer C A and Samet H, 1986 "An Optimal Construction Algorithm" in Proc 8th Int Conf on Pattern Recognition, Paris, p 317

Tanimoto S L, 1983 "A Pyramidal Approach to parallel Processing" in Proc 10th Int Symp on Computer Architecture, Stockholm, p 372

Uhr L, 1981 "Converging Pyramids of Arrays" in Proc IEEE Computer Society Workshop on Computer Architecture for Pattern Analysis and Image data base management, Hotsprings, U S A, p 1981

Chapter 12

OPTICAL RESONATORS

R K THARÉJĀ

12.1 Propagation of gaussian beams

The function of optical resonators in laser physics is similar to that of their counterparts in the microwave region. Their primary aim is twofold: (*i*) to build up large field intensities at specified (*i. e.* resonance) frequencies with moderate power inputs, and (*ii*) to act as spatial and frequency filters responding selectively to fields with prescribed spatial variation and frequency. The essential difference between an optical resonator and a microwave resonator is that in the microwave region it is possible to limit the resonances by having resonator dimensions equal to a typical wavelength, say 1 cm, but in the optical region, with wavelengths of 10^{-4} cm, it is not possible to have a resonator of the dimension of the wavelength. In general, a laser resonator consists of two mirrors. The simplest and most commonly employed resonator is a plane parallel Fabry-Pérot interferometer. Laser beams are in many respects similar to plane waves. However, their intensity distribution is not uniform, they have finite extent in the transverse direction and a spread with the distance z of travel, the phase fronts being slightly curved. Usually the intensity distribution at planes normal to the propagation direction in gaussian. The propagation, of course, obeys the Maxwell equations. If we restrict to beams which have small divergence and to media the refractive indices of which do not vary very appreciably in the transverse direction the wave equation can be reduced to a scalar one.

Consider a plane wavefront of a monochromatic electromagnetic field in the xy-plane having a gaussian distribution of the field given by

$$E(x, y) = A\, e^{-\frac{x^2 + y^2}{\omega_0^2}}\, e^{-jkz} \qquad 12.1$$

As usual, $k = (\omega/c)n = 2\pi/\lambda$, with $\lambda = \lambda_{vac}/n$, is the magnitude of the wavevector, n the refractive index of the medium and λ the wavelength in the medium. ω_0 is called the *spot radius*, defined as the distance from the centre of the gaussian distribution to a point where the field drops to 1/e times that at the centre. We shall use the scalar theory of diffraction developed in Chapter 2 to calculate the diffraction of a gaussian distribution given by eq 12.1. Following eq 2.18, the complex amplitude $U(x,y,z)$ at some arbitrary point P_2 of a monochromatic electromagnetic field is related to $U(x_0, y_0, z = 0)$, *viz* that on an aperture Σ, by the Fresnel-Kirchhoff integral

$$U(x, y, z) = \frac{jk}{2\pi} \iint_{\Sigma} U(x, y, 0) \frac{e^{-jk|\boldsymbol{r} - \boldsymbol{r}_0|}}{|\boldsymbol{r} - \boldsymbol{r}_0|} K(\theta)\, dS \qquad 12.2$$

where the integration is over the entire aperture of the field $U(x, y, 0)$ and dS is an areal element in the aperture plane $z = 0$. Eq 12.2 can be simplified by making the following approximations:

(*a*) The optical rays of interest travel close to the z-axis, the angle θ is very small so the obliquity factor is close to unity.

(*b*) The distance z from the observation plane to the aperture plane is much greater than its transverse dimensions (x_0, y_0) and (x, y) so we can replace $|\mathbf{r} - \mathbf{r}_0|$ by z in the denominator; however, we cannot do the same in the exponent.

Eq 12.2 may then be rewritten as

$$U(x, y, z) = \frac{jk}{2\pi}\frac{e^{-jkz}}{z} \iint_{\Sigma} U(x_0, y_0, z = 0)\, e^{\frac{-jk}{2z}[(x - x_0)^2 + (y - y_0)^2]}\, dx_0\, dy_0 \qquad 12.3$$

to arrive at which we have expanded the optical distance $|\mathbf{r} - \mathbf{r}_0|$

$$|\boldsymbol{r} - \boldsymbol{r}_0| = \left[(x - x_0)^2 + (y - y_0)^2 + z^2\right]^{\frac{1}{2}} \qquad 12.4$$

in a Taylor series and retained terms up to the second order in x/z, x_0/z, y/z, y_0/z only. If the field distribution at $z = 0$ is assumed to be a gaussian one given by

$$U(x_0, y_0, z = 0) = A\, e^{-\frac{x_0^2 + y_0^2}{\omega_0^2}} \qquad 12.5$$

it follows from eq 12.3 that

$$U(x, y, z)$$

$$= \frac{jkA}{2\pi}\frac{e^{-jkz}}{z} \int_{-\infty}^{\infty}\int_{-\infty}^{\infty} \exp\left[-\frac{x_0^2 + y_0^2}{\omega_0^2} - jk\frac{(x - x_0)^2 + (y - y_0)^2}{2z}\right] dx_0\, dy_0 \qquad 12.6$$

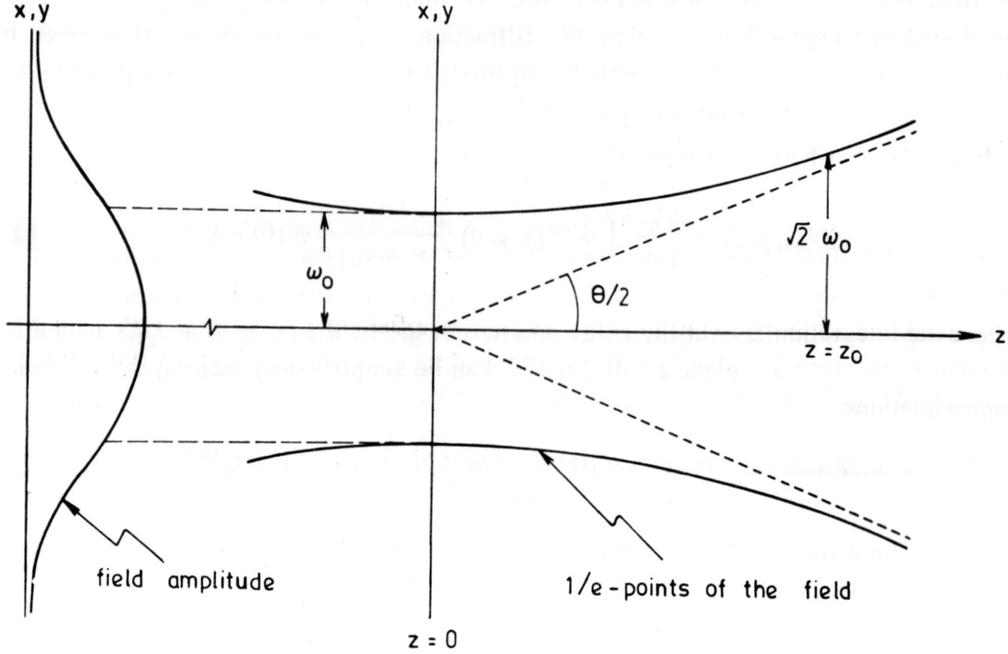

Figure 12.1 Propagating gaussian beam

Integrating this we get

$$U(x, y, z)$$

$$= \frac{jkA}{z} \frac{2\,\omega_0^2}{2z + jk\omega_0^2} \exp\left[-jk\left(z + \frac{x^2 + y^2}{2R(z)}\right)\right] \exp\left[-\frac{x^2 + y^2}{\omega^2(z)}\right] \qquad 12.7$$

We write this as

$$U(x, y, z) = \left\{A \frac{\omega_0}{\omega(z)} e^{-\frac{r^2}{\omega^2(z)}}\right\} \{e^{-j[kz - \phi(z)]}\} \left\{e^{-j\frac{kr^2}{2R(z)}}\right\} \qquad 12.8$$

where the following abbreviations have been used:

$$r \equiv (x^2 + y^2)^{\frac{1}{2}}$$

$$\phi(z) \equiv \tan^{-1}\left(\frac{z}{z_0}\right), \; z_0 \equiv \frac{\pi\,\omega_0^2}{\lambda} \qquad 12.9$$

$$\omega^2(z) \equiv \omega_0^2 \left[1 + \left(\frac{\lambda z}{\pi \omega_0^2} \right)^2 \right] \tag{12.10}$$

$$R(z) \equiv z \left[1 + \frac{\pi \omega_0^2}{\lambda z} \right] \tag{12.11}$$

In eq 12.8 the first factor describes the amplitude of the field as a function of the radial coordinate. It describes the change in amplitude as the beam propagates along z. The variation of the field with z ensures conservation of power in the beam. The field decreases to 1/e at $r = \omega$ from its value at $r = 0$. The second factor expresses the longitudinal phase, *i. e.* the change in the phase of the wave in the direction of propagation. The last factor defines the radial phase. It is obvious from eq 12.8 that the field distribution in any plane perpendicular to the z-axis remains gaussian with the spot radius being defined by eq 12.10. Hence a beam that has an initial gaussian intensity distribution remains gaussian as it propagates. Figure 12.1 (p. 178) shows a propagating gaussian beam.

12.2 Propagation of a gaussian beam

(*a*) *Spot size* The spot size is defined by eq 12.10 and we have

$$\omega^2 = \omega_0^2 \left[1 + \left(\frac{\lambda z}{\pi \omega_0^2} \right)^2 \right] = \omega_0^2 \left[1 + \frac{z^2}{z_0^2} \right]$$

$$\frac{\omega^2}{\omega_0^2} - \frac{z^2}{z_0^2} = 1 \tag{12.12}$$

This represents a hyperbola as shown in Figure 12.1. This shows that the beam spreads out as it travels away from the origin. Since z occurs in an even power in eq 12.10 the beam width increases along both positive and negative z-directions. The width is minimum at $z = 0$ and the phase front is plane. $[\omega]_{z=0} \equiv \omega_0$ is called the *waist of the gaussian beam.*

(*b*) *Beam divergence* For large enough z eq 12.10 becomes

$$[\omega(z)]_{z \gg z_0} = \frac{\omega_0 z}{z_0} = \frac{\lambda z}{\pi \omega_0} \tag{12.13}$$

The divergence angle of the beam is defined as

$$\theta = \frac{d\omega}{dz} = \frac{\lambda}{\pi \omega_0} \tag{12.14}$$

This is the minimum spread a beam of diameter $2\omega_0$ can have. One can define beam divergence θ as half the apex angle of the cone formed by all of the asymptotes only two of which are shown in Figure 12.1. We know that the diffraction angle corresponding to the Airy disk of a circular aperture of radius ω_0 is $\theta_d = 1.22\lambda / \omega_0$. This angle is greater than the beam divergence angle in eq 12.14. The diffraction losses of a gaussian beam are therefore smaller than those for a beam with uniform intensity across.

From eq 12.10-11 it follows that at $z = z_0$ we have $\omega^2 = 2\omega_0^2$ and $R(z) = 2z_0$. Thus at $z = z_0$ the beam diameter is double that at the centre. In the region between $z = \pm z_0$ the irradiance on the axis is more than half of its maximum value. Hence $2z_0$ is generally taken as the depth of focus of the gaussian beam.

(c) *Curvature of wavefronts* Wavefronts are surfaces of constant phase and can be obtained by equating the imaginary part of the exponent in eq 12.8 to a constant. The wavefront passing through the point $(0, 0, z')$ is given by

$$k\left[z + \frac{x^2 + y^2}{2R(z)}\right] - \tan^{-1}\left(\frac{z}{z_0}\right) = kz' - \tan^{-1}\left(\frac{z'}{z_0}\right) \qquad 12.15$$

Since $z \approx z'$, $\tan^{-1}(z/z_0) \approx \tan^{-1}(z'/z_0)$, eq 12.15 reads

$$z + \frac{x^2 + y^2}{2R(z)} = z'$$

$$z' - z = \frac{x^2 + y^2}{2R(z)} \qquad 12.16$$

This represents a spherical surface of radius of curvature R, defined by eq 12.11. From the latter it follows that only when $z \gg z_0$ the beam appears to originate at $z = 0$. As one gets closer to that point the centre recedes until at $z = 0$ the centre of curvature is at infinity and the wavefront becomes planar.

Summarizing these results, a gaussian beam is completely characterized by parameters ω and R in any given plane. At $z = 0$ we have $\omega = \omega_0$ and $R = \infty$.

12.3 Transformation of gaussian beams

In eq 12.8 define a complex parameter q

$$\frac{1}{q} = \frac{1}{R} - \frac{j\lambda}{\pi\omega^2} \qquad 12.17$$

and write that equation in terms of q as

$$\omega(x, y, z) = A\frac{\omega_0 e^{-\frac{r^2}{\omega^2}}}{\omega(z)} e^{-j\left[kz - \tan^{-1}\left(\frac{z}{z_0}\right) + \frac{kr^2}{2q(z)}\right]} \quad 12.18$$

Eq 12.18 and 12.8 have similar dependence on the various parameters; q may therefore be termed the *complex radius of curvature* of the spherical gaussian beam. From eq 12.10 and 12.17 we may write

$$q(z) = z + j\frac{\pi\omega_0^2}{\lambda} = z + jz_0 \quad 12.19$$

$$\equiv z + q_0,\ q_0 \equiv jz_0 = j\frac{\pi\omega_0^2}{\lambda} \quad 12.20$$

It follows from eq 12.20 that if a beam propagates from distance $z = z_1$ to $z = z_2$ the radii of curvature at the two planes satisfy the relation

$$q_2(z_2) = q_1(z_1) + d,\ d = z_2 - z_1 \quad 12.21$$

To work with any beams in experiments we use lenses and other refracting devices for transforming them. Consider the gaussian beam, shown in Figure 12.2 incident from vacuum on to a transparent medium of refractive index n. As the beam crosses the boundary the radius of curvature of the wavefront changes from R to R' (= nR), the beam radius remains unchanged, but $z' = nz$ and $z_0' = nz_0$ so the location of the beam changes. Although the beam waist radius remains the same the beam divergence in the medium of refractive index n is $1/n$ times that of the beam in vacuum. Thus, for a gaussian beam passing through a focussing lens with focal length f, the beam size ω remains the same on

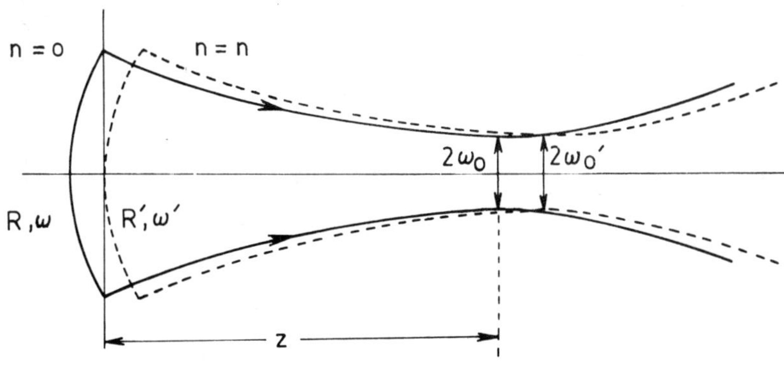

Figure 12.2 Propagation of a gaussian beam across a boundary between two media

both sides of the lens and the radius of curvature R_1 changes to R_2 in the same way as for spherical waves:

$$\frac{1}{R_2} = \frac{1}{R_1} - \frac{1}{f} \tag{12.22}$$

In terms of q eq 12.22 reads

$$\frac{1}{q_2} = \frac{1}{q_1} - \frac{1}{f} \tag{12.23}$$

Consider a case in which a beam is transmitted through a lens as shown in Figure 12.3. The beam propagates in the following segments:

(*a*) The beam is first transformed in free space between P_1 and the lens so

$$q_2 = q_1 + d_1 \tag{12.24}$$

(*b*) It is then transformed in the lens so

$$\frac{1}{q_3} = \frac{1}{q_1} - \frac{1}{f} \tag{12.25}$$

(*c*) Finally it is transformed in free space between the lens and P_2 so

$$q_4 = q_3 + d \tag{12.26}$$

From these three relations we get q_4 as

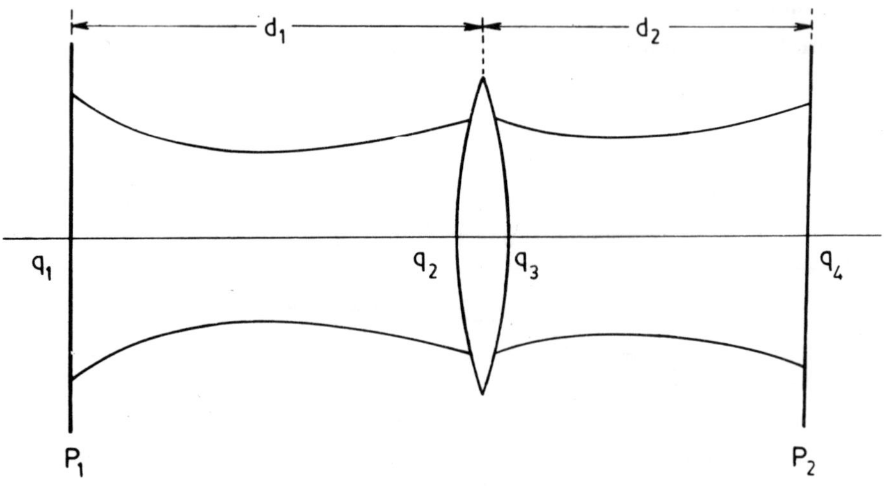

Figure 12.3 Propagation of a gaussian beam through a lens

$$q_4 = \frac{\left(1 - \frac{d_2}{f}\right) q_1 + \left(d_1 + d_2 - \frac{d_1 d_2}{f}\right)}{\left(1 - \frac{d_1}{f}\right) - \frac{q_1}{f}} \qquad 12.27$$

Using this equation one can easily find the spot size ω and the radius of curvature R at any distance d_2 behind the lens. Let us demonstrate the use of eq 12.27 in an example in which a laser beam is focussed using the lens. We wish to transform the beam waist of the laser resonator into a beam waist located in the interaction region. The beam parameters in the waist are purely imaginary, *viz*

$$q_1 = j\frac{\pi \omega_1^2}{\lambda}, \quad q_2 = j\frac{\pi \omega_2^2}{\lambda} \qquad 12.28$$

and the radius of curvature is infinite. From eq 12.27 and 12.28, equating the real and imaginary parts respectively and noting that $q_4 \equiv q_2$ we get

$$\frac{d_1 - f}{d_2 - f} = \frac{\omega_1^2}{\omega_2^2} \qquad 12.29$$

$$\omega_2 = \omega_1 \left[\frac{d_2 - f}{d_1 - f}\right]^{1/2} \qquad 12.30$$

and

$$(d_1 - f)(d_2 - f) = f^2 - \frac{\pi \omega_1 \omega_2}{\lambda}$$

$$= f^2 - f_0^2 \qquad 12.31$$

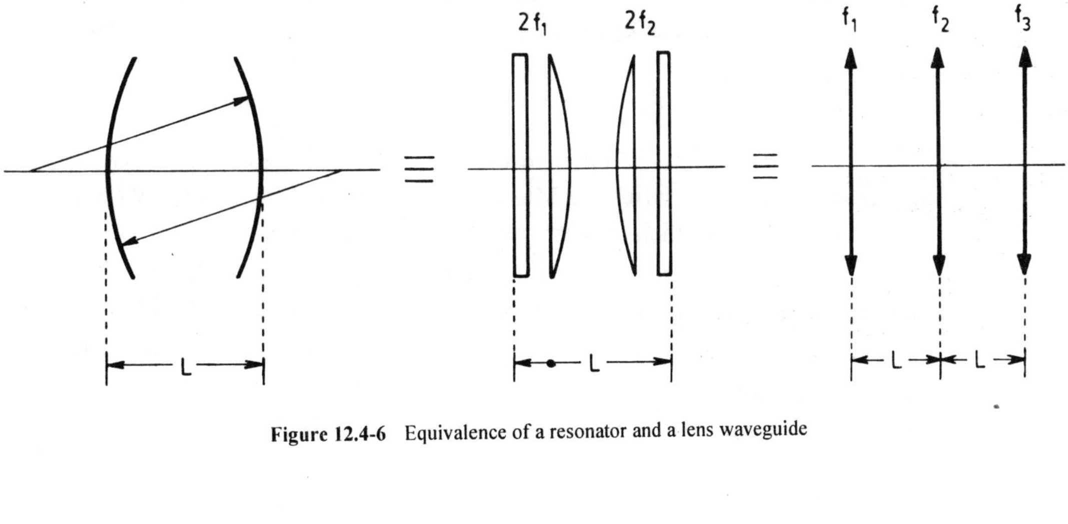

Figure 12.4-6 Equivalence of a resonator and a lens waveguide

Since d_1, $d_2 > f$ this shows that any lens with $f > f_0$ can be used. We can find the position of the lens of given f by solving for d_1 and d_2:

$$d_1 = f \pm \frac{\omega_1}{\omega_2}\left[f^2 - f_0^2\right]^{1/2}$$

$$d_2 = f \pm \frac{\omega_2}{\omega_1}\left[f^2 - f_0^2\right]^{1/2} \qquad 12.32$$

In all cases where it is required to introduce a gaussian beam emerging from one resonator to another it is important that the initial beam be properly transformed or matched with the receiving resonator. To do this, one usually uses a lens at a definite point between the first and second resonators.

12.4 Optical resonators and Lens waveguides

In an optical cavity a quasiplanar optical wave bounces back and forth between the two mirrors. The system is equivalent to a long waveguide, or optical transmission line, in which the light beam is transmitted with little or no loss through a repeated series of identical lenses mounted in a line. Consider a resonator of length L formed by two concave spherical mirrors with radii of curvature R_1 and R_2. The resonator shown in Figure 12.4 (*see* last page) is equivalent to the resonator shown in Figure 12.5 (*see* last page) which uses combinations of a plane mirror and planoconvex lens instead of concave mirrors. This in turn is optically equivalent to a lens waveguide consisting of double convex lenses with focal lengths f_1 and f_2 alternatingly, the distance between adjacent lenses being L. Hence Figure 12.5 is equivalent to Figure 12.6 (*see* last page). Of course, the light wave changes the direction of its propagation in systems shown in Figures 12.4 and 12.5 each time it reaches a mirror while in the system of Figure 12.6 the light wave always propagates in the same direction. The difference between a lens waveguide and a resonator is that while the resonator isolates certain frequencies the waveguide has a continuous frequency spectrum. In all other respects a resonator and a waveguide are alike: certain transverse modes are excited and maintained in a waveguide just as in the case of the resonator.

A mode of a cavity (that is, an open resonator) is a sourceless configuration of the electromagnetic field which satisfies the Maxwell equations and the boundary conditions imposed by the mirrors, and which oscillates with a slowly decaying amplitude as $U(x, y, z)\, exp(j\omega t - t/\tau)$. The decay, defined by the time constant τ, is due to the leaking out of the energy from the open structure, if mirror losses like absorption are negligible. In actual practice the energy supplied by stimulated emission will compensate for all losses and secure steady state oscillation. A rigorous treatment of a plane parallel resonator is due to Fox and Li. They used the scalar theory of diffraction to analyze the problem. In this approximation the electromagnetic field is described by a scalar function of the spatial coordinates assumed to be nearly transverse and uniformly polarized (§ 2.1). We shall use eq 12.3 to evaluate the complex amplitude and phase $v(x,y)$ of the wave at a plane $z_0 + L = z$ from the given initial complex amplitude $u_0(x_0, y_0)$, L being the length of the plane parallel resonator.

If $u_0(x_0, y_0)$ is an arbitrary field at the left resonator mirror we can calculate the field $v'(x, y)$ at the right mirror using eq 12.3. Now using $v(x,y)$ in eq 12.3 we can find the new value of u_0 which can be used to find the new value of v' and so on. Fox and Li showed that, after a large number of passages in the resonator from one mirror to the other, the structure of the field or its dependence on the transverse coordinates starts getting reproduced, repeating itself after every passage of the radiation through the resonator. This means that after a large number of transitions of the field at the surface of the mirror is described by the function u_0 which is a solution of the Fox-Li integral equation

$$\frac{jk\,e^{-jkL}}{2\pi}\iint_{\substack{input\\plane}} u_0(x_0, y_0)\, e^{\frac{-jk}{2L}[(x-x_0)^2+(y-y_0)^2]}\,dx_0\,dy_0 = \gamma\, u_0(x, y) \qquad 12.33$$

Irrespective of the transverse coordinates the complex factor takes into account the phase shift in the field for a passage (including the purely geometrical phase incursion over a distance L, given by the exponential factor γ) as well as the decrease in the amplitude u_0 of the field as a result of losses in the resonator.

Eq 12.33 may be rewritten as

$$v(x, y) = \frac{jk\,e^{-jkL}}{2\pi}\iint_{\substack{input\\plane}} u_0(x_0, y_0)\,K(x_0, y_0, x, y)\,dx_0\,dy_0 \qquad 12.34$$

where the function K is defined as

$$K(x_0, y_0, x, y) \equiv e^{\frac{-jk}{2L}[(x-x_0)^2+(y-y_0)^2]} \qquad 12.35$$

K is usually called the *propagation kernel*. If instead of plane mirrors we use two rectangular mirrors with radii of curvature R_1 and R_2 to form an optical resonator of length L the kernel will be given by

$$K(x_0, y_0, x, y) = e^{\frac{-jk}{2L}[g_1(x_0^2+y_0^2)+g_2(x_1^2+y_1^2)-2(x_0y_0+x_1y_1)]} \qquad 12.36$$

where the g's are defined as

$$g_{1,2} = 1 - \frac{L}{R_{1,2}} \qquad 12.37$$

Eq 12.33 may be interpreted as an equation with an integral operator describing the transformation of the field in the resonator for one pass, γ being the eigenvalues of the operator and u_0 its eigenfunctions. The spectrum of the eigenvalues of the operator is discrete. In order to fix the various eigenvalues we make use of indices *m* and *n*, each being 0,1,2,3,... The eigenfunction corresponding to the eigenvalue γ_{mn} is denoted by u_{mn}. This is the *mn*th transverse mode for a given resonator. Knowing the eigenvalues γ_{mn} we can find the losses for the *mn*th mode, namely $(1 - |\gamma_{mn}|^2)$. The phase shift of the *mn*th mode is given by arg γ_{mn}.

12.5 Resonance frequencies of the modes

The spectrum of the resonance frequencies for the transverse modes can be found by equating the phase shift for a pass through the resonator to the product of π and an integer *p*, *i. e.* $2kL = 2p\pi$ or $L = p\lambda/2$. In terms of the expression for the axial phase shift for a gaussian beam (*see* eq 12.18) we have

$$\phi(z_2) - \phi(z_1) = p\pi \qquad 12.38$$

Since

$$\phi(z) = kz - (m+n+1)\tan^{-1}(z/z_0)$$

we get

$$p\pi = \frac{2\pi\nu}{c} - (m+n+1)\left[\tan^{-1}(z_2/z_0) - \tan^{-1}(z_1/z_0)\right]$$

The frequencies are then given by

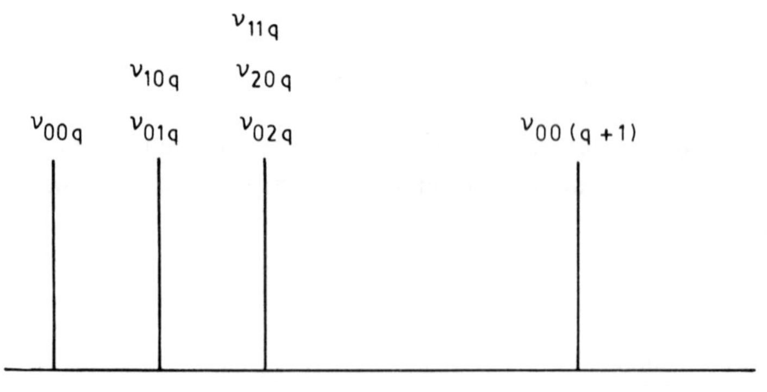

Figure 12.7 Frequency spectrum of a resonator

$$\nu_{mnp} = \left[p + \frac{m+n+1}{\pi} \left\{ \tan^{-1}(z_2/z_0) - \tan^{-1}(z_1/z_0) \right\} \right] \frac{c}{2L}$$

Substituting for z_1 and z_2 in terms of g_1 and g_2 we obtain

$$\nu_{mnp} = \left[p + \frac{m+n+1}{\pi} \cos^{-1}\sqrt{g_1 g_2} \right] \frac{c}{2L} \qquad 12.39$$

Thus the resonant frequency is degenerate in *m* and *n*; in other words, for all combinations of m and n for which m + n = constant the resonance frequency is the same (Figure 12.7, p. 186). For a confocal resonator, *i. e.* for $L = R_1 = R_2$, we get an additional degeneracy. This property is used in the scanning spherical mirror Fabry-Pérot interferometer.

12.6 Higher order modes

Eq 12.34 is usually solved numerically. Assuming that the mirrors have quadratic aperture, cartesian coordinates can be used to evaluate it. In rectangular coordinates its eigenfunctions are

$u_{mn}(x, y, z)$

$$= \left[\frac{1}{2^{m+n}\, m!\, n!} \right]^{\frac{1}{2}} \frac{1}{\omega(z)} H_m\left(\frac{\sqrt{2}\,x}{\omega(z)} \right) H_n\left(\frac{\sqrt{2}\,y}{\omega(z)} \right) e^{\frac{jk(x^2+y^2)/2}{q(z)}} e^{-jkz + j(m+n+1)\phi(z)} \qquad 12.40$$

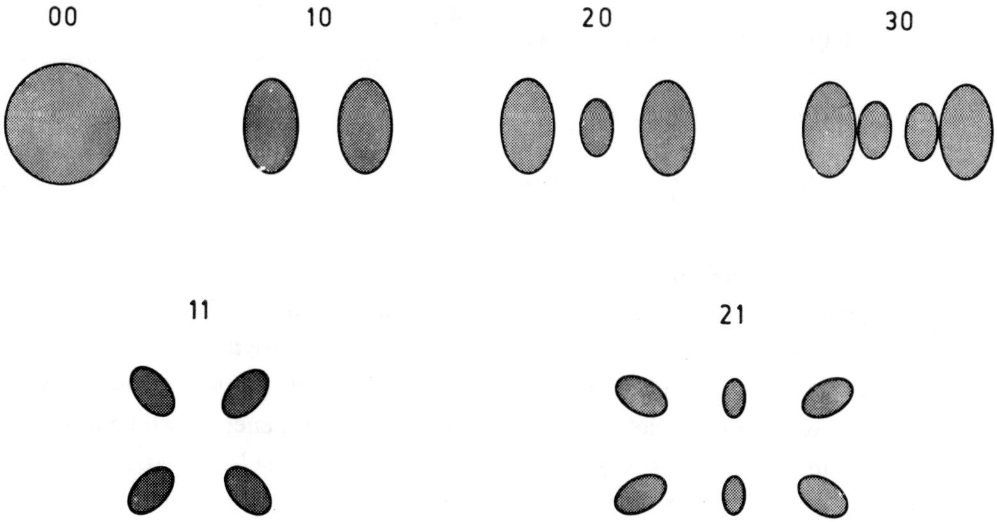

Figure 12.8 Mode patterns

where $\phi(z)$, $\omega(z)$ and $q(z)$ have already been defined in the text above. $H_n(z)$ are Hermite polynomials defined as

$$H_n(z) = (-1)^n e^{z^2} \frac{\partial^n}{\partial z^n} e^{-z^2} \tag{12.41}$$

Eq 12.40 states that the irradiance pattern in a cross-section of a higher order beam is described by the product of Hermite and gaussian functions. Each of the higher order transverse modes retains the same transverse form at every longitudinal position z, except for a change in transverse scaling due to the factor $\omega(z)$ accompanying every factor x or y. A photograph of a mode pattern will essentially follow the properties of the Hermite polynomials: (1) even numbered polynomials are symmetric and odd numbered antisymmetric, (2) the nth order polynomial contains n zeros or nulls, and $(n-1)$ peaks or lobes, the outermost lobe being the strongest. In a photograph each lobe will appear as an exposed spot. The m, n values of a spatial mode can be found by counting the number of dark bars across the pattern in the x and y directions. A few mode patterns are shown in Figure 12.8 (p. 187). Higher order gaussian modes may also be written in circular polar coordinates in the Gaussian-Laguerre form. A stable optical resonator will give higher order gaussian beams as well as the lowest order gaussian beams, hence the system will oscillate in many modes simultaneously. Since the wavefront curvature $R(z)$ and spot size $\omega(z)$ are independent of the mode indices m and n a laser oscillating in various modes will have its energy distributed in all of the modes present. However, the phase shift is a function of the mode number and is given by

$$\phi(m, n, z) = (m + n + 1) \tan^{-1}\left(\frac{\lambda z}{\pi \omega_0^2}\right) \tag{12.42}$$

$$\phi(p, l, z) = (2p + l + 1) \tan^{-1}\left(\frac{\lambda z}{\pi \omega_0^2}\right) \tag{12.43}$$

for cartesian and cylindrical coordinate systems respectively. To have a single mode of oscillation we must find a way of favouring the lowest order mode. The easiest way to do so is to make the gains maximum on the axis, or somehow increase the losses for the higher order modes. If we place an aperture of diameter $2a$ in the cavity all of the mode losses will increase as a decreases to mode size ω because more and more energy will be lost around the edges. Higher order modes which have large transverse spread will suffer the greatest losses. In this way the higher order modes will be suppressed. The optimum value for the aperture diameter is found to be $2a = 3.5 - 4.0\ \omega$.

12.7 Gaussian beams in stable resonators

As discussed earlier, in a resonator the lowest order resonant or cavity mode is a gaussian beam of appropriate curvature and spot size trapped between two mirrors. However, to analyze a system it is best to start backwards, *viz* to assume that a gaussian beam is propagating between two mirrors and fit the resonator mirrors spacing, *etc*. Thus we start by assuming a gaussian beam having a waist of spot size ω_0 at $z = 0$ and spot sizes ω_1 and ω_2, radii of curvature R_1 and R_2, at two arbitrary planes $z = z_1$ and $z = z_2$ as shown in Figure 12.9. If we place a mirror at points z_1 and z_2 each, having radii of curvature R_1 and R_2 and with sizes greater than ω_1 and ω_2 respectively, the gaussian beam will travel back and forth between the two mirrors forming a standing wave without its spot size or curvatures varying with time. Our aim is to find spot sizes ω_1, ω_2, beam waist ω_0, waist location, *etc* in this configuration. We have

$$R(z) = z + \frac{1}{z}\left(\frac{\pi\,\omega_0^2}{\lambda}\right)^2 = z + \frac{z_0^2}{z} \qquad 12.44$$

where $z_0 \equiv \pi\,\omega_0^2/\lambda$, the *Rayleigh length*. From Figure 12.9 we have

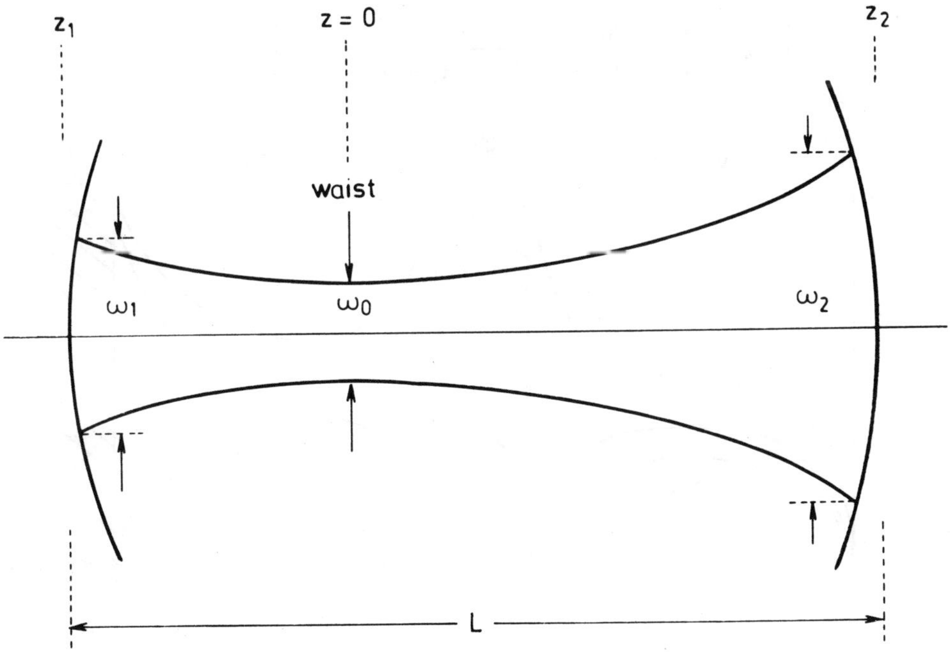

Figure 12.9 Laser resonator configuration

$$-R_1 = z_1 + \frac{z_0^2}{z_1}, \quad R_2 = z_2 + \frac{z_0^2}{z_2}, \quad z_2 - z_1 = L \tag{12.45}$$

Solving for z_1 and z_2 we get

$$z_2 = \frac{L\,(R_1 - L)}{R_1 + R_2 - 2\,L}, \quad z_1 = \frac{L\,(R_2 - L)}{R_1 + R_2 - 2\,L}$$

$$z_0^2 = \frac{L\,(R_1 - L)\,(R_2 - L)\,(R_1 + R_2 - L)}{(R_1 + R_2 - 2\,L)^2} \tag{12.46}$$

Since

$$= \omega_1^2 = \omega_0^2\,[1 + (z_1 / z_0)^2]$$

we have

$$\omega_1^4 = \left(\frac{\lambda R_1}{\pi}\right)^2 \frac{R_2 - L}{R_1 - L}\,\frac{L}{R_1 + R_2 - L} \tag{12.47a}$$

$$\omega_2^4 = \left(\frac{\lambda R_2}{\pi}\right)^2 \frac{R_1 - L}{R_2 - L}\,\frac{L}{R_1 + R_2 - L} \tag{12.47b}$$

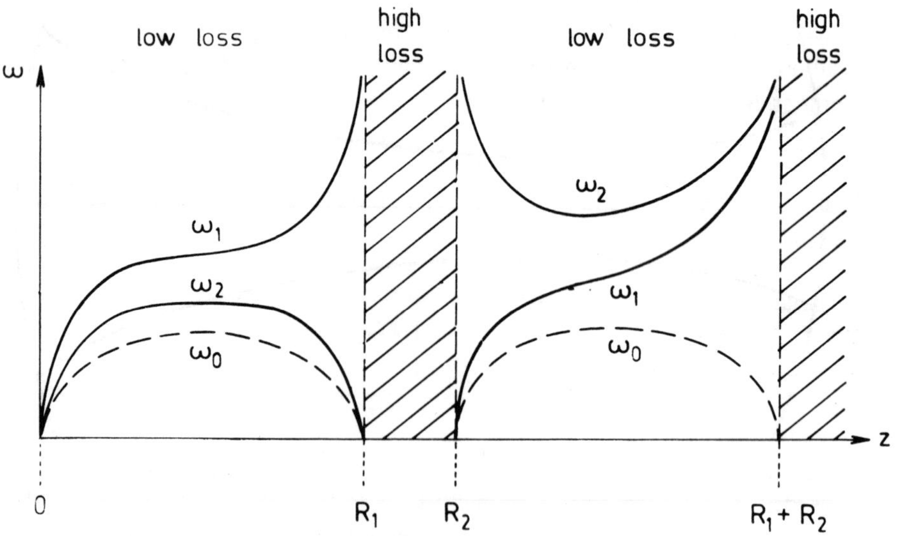

Figure 12.10 Spot size variation and losses

$$\omega_0^4 = \left(\frac{\lambda}{\pi}\right)^2 \frac{L\,(R_1 - L)\,(R_2 - L)\,(R_1 + R_2 - L)}{(R_1 + R_2 - 2\,L)^2} \qquad 12.47c$$

It is interesting to note that ω_0, ω_1 and ω_2 are independent of the sizes of the mirrors. All of these spot sizes are shown in Figure 12.10 (p. 190).

As discussed in the text, z_0 defines a scale length and is sufficient to define a beam in a resonator. For a system to be stable, *i. e.* to sustain a gaussian beam, z_0 must be real. Thus we have the following four regions, depending on R_1, R_2 and L:

(a) $R_1 > L$ and $R_1+R_2 > L > R_2$: z_0 is real; the resonator has low loss and is stable.

(b) $R_2 > L > R_1$ and $L > R_1+R_2$: z_0 is imaginary; the resonator cannot sustain a gaussian beam and is unstable.

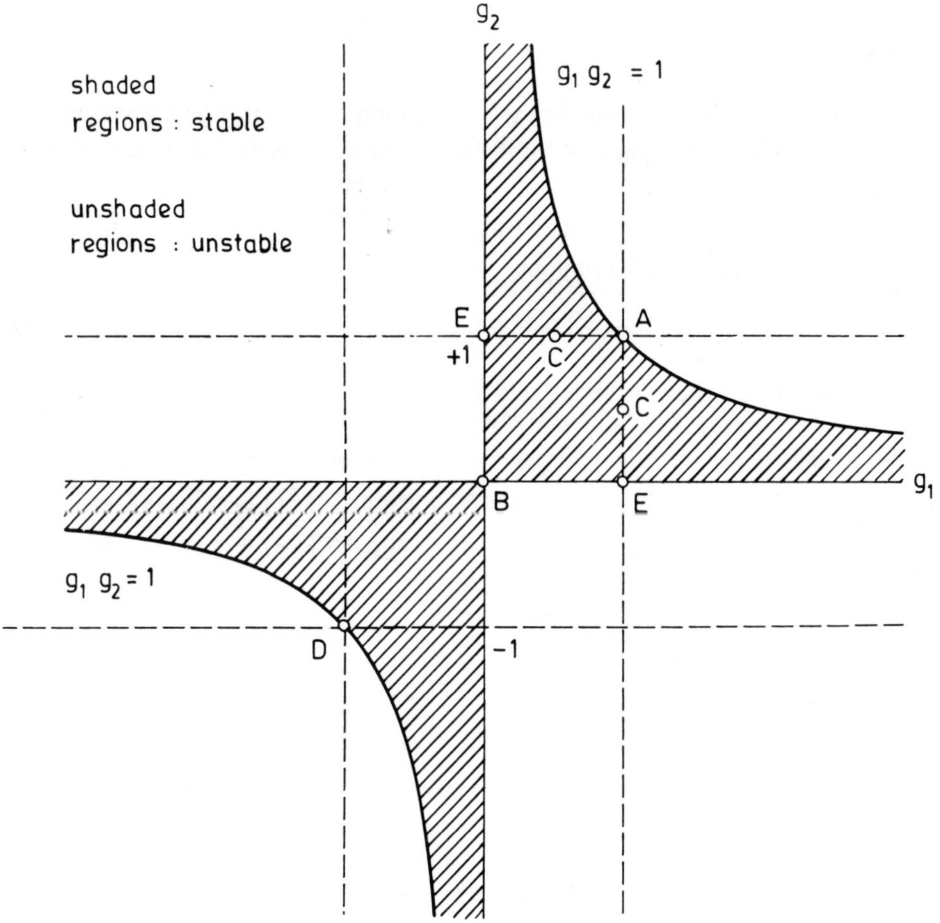

Figure 12.11 Stability diagram

It is possible to show on a single graph whether a resonator is stable or unstable by taking g_1 and g_2 as the coordinate axes (Figure 12.11, last page). Each resonator system is represented by a point in this diagram. The boundary lines between the stable and unstable regions are two straight lines given by

$$L / R_1 = 1, \; L / R_2 = 1 \qquad 12.48$$

and a hyperbola which satisfies

$$L = R_1 + R_2 \;\; or \;\; g_1 \, g_2 = 1 \qquad 12.49$$

The origin of the diagram represents the confocal system. In actual practice it is preferable deliberately to deviate a bit towards the stability region by changing the length of the resonator. This is not possible in the case of a plane parallel resonator.

12.8 Matrix method

We make the paraxial approximation for the propagation of light in a lens waveguide or in a resonator. A paraxial ray propagating in the z-direction at a certain distance z is characterized by two parameters: the distance $y(z)$ from the optical axis and the tangent of the angle of inclination $\alpha(z)$ to the axis. The transformation of the ray as it propagates from $z = z_1$ to $z = z_2$ is described by the relations

$$y_2 = A \, y_1 + B \, \alpha_1, \alpha_2 = C y_1 + D \, \alpha_1 \qquad 12.50$$

In matrix form one writes the above relations as

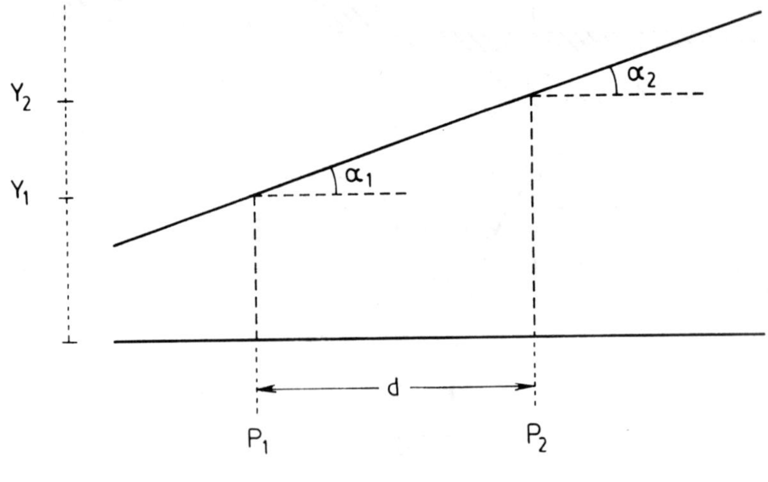

Figure 12.12 Ray passage through free space

$$\begin{bmatrix} y_2 \\ \alpha_2 \end{bmatrix} = \begin{bmatrix} A & B \\ C & D \end{bmatrix} \begin{bmatrix} y_1 \\ \alpha_1 \end{bmatrix} \qquad 12.51$$

The matrix [A B C D] is called the *transformation matrix*. Suppose that a ray passes through free space extending over length d (Figure 12.12, p. 192). Then we have

$$y_2 = y_1 + \alpha_1 d, \; \alpha_2 = \alpha_1$$

$$A = 1, \; B = d, \; C = 0, \; D = 1 \qquad 12.52$$

$$M(d) = \begin{bmatrix} A & B \\ C & D \end{bmatrix} = \begin{bmatrix} 1 & d \\ 0 & 1 \end{bmatrix} \qquad 12.53$$

The transformation of a light ray in a thin converging lens of focal length f can be written as

$$M(f) = \begin{bmatrix} 1 & 0 \\ -1/f & 1 \end{bmatrix} \qquad 12.54$$

Similarly the transformation matrix for a ray being reflected from a mirror with radius of curvature r is given by

$$M(r) = \begin{bmatrix} 1 & 0 \\ -2/r & 1 \end{bmatrix} \qquad 12.55$$

The transformation of a ray upon successive passage through free space of distance d, then through a thin lens of focal length f, can be written as the product of two matrices:

$$\begin{bmatrix} 1 & 0 \\ -1/f & 1 \end{bmatrix} \begin{bmatrix} 1 & d \\ 0 & 1 \end{bmatrix} = \begin{bmatrix} 1 & d \\ -1/f & 1-d/f \end{bmatrix} \qquad 12.56$$

Consider a resonator of length L with mirrors having radii of curvature R_1 (left mirror) and R_2 (right mirror). Choose the reference plane P at a distance d from the left mirror (Figure 12.13a, p. 194). The transformation matrix of the ray for a round trip through the resonator from the plane P can be written as

$$M[d \to R_1 \to L \to R_2 \to (l-d)] = M[l-d]\, M[R_2]\, M(L)\, M(R_1)\, M(d) \qquad 12.57$$

where we have

$$M(L-d) = \begin{bmatrix} 1 & l-d \\ 0 & 1 \end{bmatrix}, \; M(R_2) = \begin{bmatrix} 1 & 0 \\ -2/R_2 & 1 \end{bmatrix}, \; M(L) = \begin{bmatrix} 1 & L \\ 0 & 1 \end{bmatrix}$$

$$M(R_1) = \begin{bmatrix} 1 & 0 \\ -2/R_1 & 1 \end{bmatrix}, \; M(d) = \begin{bmatrix} 1 & d \\ 0 & 1 \end{bmatrix} \qquad 12.58$$

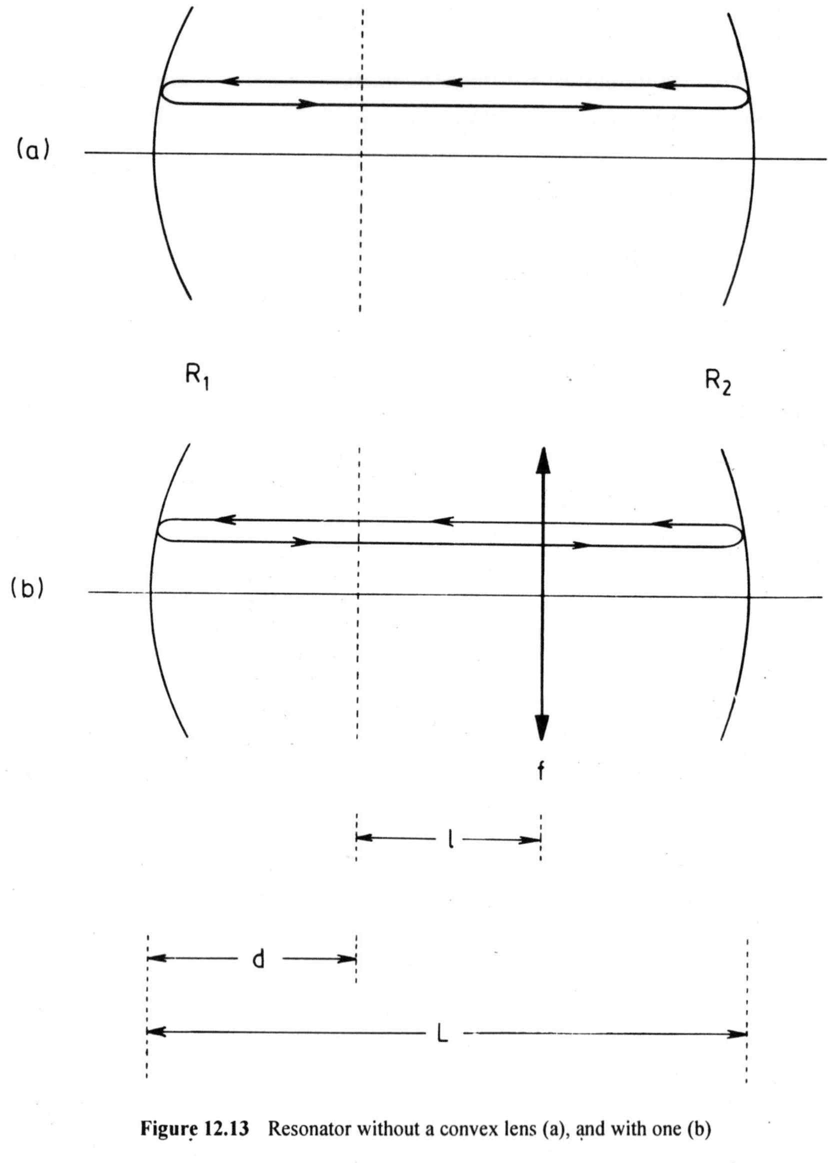

Figure 12.13 Resonator without a convex lens (a), and with one (b)

12.9 Unstable resonators for high power lasers

We have seen that the stability criterion for a stable resonator is $0 < g_1 g_2 < 1$. Resonators for which this condition is not satisfied are called unstable resonators. Unstable resonators can be classified into two classes: (*i*) positive branch resonators corresponding to $g_1 g_2 > 1$ and (*ii*) negative branch resonators corresponding to $g_1 g_2 < 0$. For the case of a stable configuration we have seen that the dimensions of the beam spot size and mode volume are small and the beam propagating back and forth between the two mirrors is focussed constantly. However, for an unstable resonator the spot size and hence mode volume increases as the beam bounces back and forth between the mirrors. The diffraction losses are very high and a high optimum coupling is possible provided the medium has high enough gains to compensate for the losses. An exact analysis of unstable resonators can be made by deriving diffraction losses and the field distribution modes, using the Fresnel-Kirchhoff integral. However, most of the results for unstable resonators could be obtained in the geometrical approximation.

The following applications of unstable resonators for a laser oscillator are worth mentioning:

(*i*) For selecting the transverse oscillatory modes and to reduce the angular divergence of radiation

(*ii*) For having large volume of the fundamental mode

(*iii*) For creating lasers with stringently prescribed spectra and time characteristics

Bibliography

ANAEV Yu A, 1972 Soviet J Quant Electr **1** 565

FOX A G and Li T, 1961 Bell Syst Tech J **40** 453

KOGELNIK H and Li T, 1966 J Appl Optics **5** 1550

SIEGMAN A E, 1965 Proc IEEE **53** 277

SVELTO O, *Principles of Lasers*

YARIV A, *Introduction to Quantum Electronics*

YARIV A, *Optical Electronics*

Chapter 13

NONLINEAR OPTICS

R K THARÉJĀ

13.1 Introduction

Soon after the invention of the laser it was demonstrated by Franken *et al* that the intense light flux produced nonlinear optical effects in various media. They produced a second harmonic of the ruby laser radiation by focussing a ruby laser beam into a quartz crystal. The development since then may be divided into the following four periods:

1961-65: Discovery of nonlinear effects, sum and difference frequency generation, harmonic generation, parametric amplification and oscillation, multiphonon absorption, self-focussing, stimulated scattering, *etc*

1965-69: New nonlinear effects, nonlinear spectroscopy, transient coherent phenomena, optical breakdown, *etc*

1969-75: Use of tunable lasers and ultra-short pulsed lasers to expand the use and study of nonlinear effects

1975-: Discovery and construction of nonlinear optical devices, optical bistability, optical phase conjugation, *etc*

In the present set of five lectures we shall restrict ourselves to a general introduction to nonlinear processes, optical bistability, optical phase conjugation and some practical devices based on nonlinear optics.

We start by giving a quick review of the fundamentals of nonlinear optics. We all know that all optical phenomena are governed by the Maxwell equations and the resulting wave equations. When a dielectric medium is subjected to an applied electric field **E** it becomes polarized due to the distortion of its internal charge distribution. The resultant electric dipole moment per unit volume is called the electric polarization **P**. In general, **P** is related to **E** through a field-dependent susceptibility tensor $\chi(\mathbf{E})$ as

$$\boldsymbol{P} = \varepsilon_0 \, \chi \cdot \boldsymbol{E}$$

χ is a property of the material of the medium. When the electric field is sufficiently large — as in the case of illumination by laser light — **P** is a nonlinear function of **E**. We have, somewhat symbolically,

$$\boldsymbol{P} = \varepsilon_0 \, \chi^{(1)} E + \varepsilon_0 \, \chi^{(2)} \, E \cdot E + \varepsilon_0 \, \chi^{(3)} \, E \cdot E \cdot E + \cdots$$

The first term, containing the first order susceptibility $\chi^{(1)}$, describes the familiar linear optical property of an index of refraction, and $\chi^{(2)}$, $\chi^{(3)}$, *etc* are called the quadratic, cubic, *etc* nonlinear susceptibility tensors. The latter are generally much smaller than $\chi^{(1)}$ and contribute significantly only at large electric fields. In most crystalline materials the nonlinear polarizability $\chi^{(2)}$ depends on the direction of propagation of the light wave, polarization of the electric field and the orientation of the optic axis of the crystal. It gives rise to nonlinear phenomena of second harmonic generation, sum and difference frequency mixing, optical rectification, linear electro-optic effects and parametric generation. The cubic susceptibility $\chi^{(3)}$ is responsible for third harmonic generation, quadratic electro-optic effects, two-phonon absorption, and stimulated raman, Brillouin and Rayleigh scattering.

13.2 Second order nonlinear phenomena

For second order processes we have

$$P_i^{(2)} = \sum_{j,k} \chi_{ijk}^{(2)} \, E_j \cdot E_k$$

where *i, j, k* represent *x, y, z*. Thus for the *x*-direction we have

$$P_x^{(2\omega)} = \chi_{xxx}^{(2\omega)} \, E_x^{(\omega)} \, E_x^{(\omega)} + \chi_{xyy}^{(2\omega)} \, E_y^{(\omega)} \, E_y^{(\omega)} + \chi_{xzz}^{(2\omega)} \, E_z^{(\omega)} \, E_z^{(\omega)}$$

$$+ \, 2\,\chi_{xzy}^{(2\omega)} \, E_z^{(\omega)} \, E_y^{(\omega)} + 2\,\chi_{xzx}^{(2\omega)} \, E_z^{(\omega)} \, E_x^{(\omega)} + 2\,\chi_{xxy}^{(\omega)} \, E_x^{(\omega)} \, E_y^{(\omega)}$$

Similar expressions may be written for $P_y^{(2\omega)}$ and $P_z^{(2\omega)}$. Considerations of crystal symmetry reduce the number of nonvanishing elements of $\chi_{ijk}^{(2\chi)}$ or, in certain cases, cause them to vanish altogether. Consider the nonlinear coupling of two optical fields of angular frequencies ω_1 and ω_2 at point **r**, the electric fields being defined as

$$E_j^{(\omega_1)}(t) = \mathrm{Re}\left(E_j^{\omega_1} \, e^{j\omega_1 t}\right) = \frac{1}{2}\left(E_j^{\omega_1} \, e^{j\omega_1 t} + c\,c\right)$$

$$E_k^{(\omega_2)}(t) = \mathrm{Re}\left(E_k^{\omega_2} \, e^{j\omega_2 t}\right)$$

wherein the subscripts j and k take the values x, y, z. The induced polarization at the sum or difference frequencies $\omega_3 = \omega_1 \pm \omega_2$ arises from the second order nonlinear susceptibility $\chi^{(2)}$. Taking the polarization component along the i-direction at frequency $\omega_3 = \omega_1 + \omega_2$ we get

$$P_i^{\omega_3 = \omega_1 + \omega_2}(t) = \text{Re}\,(P_i^{\omega_3} e^{j\omega_3 t}) = \text{Re}\,(\chi_{ijk}^{\omega_3 = \omega_1 + \omega_2} E_j^{\omega_1} E_k^{\omega_2} e^{j(\omega_1 + \omega_2)t})$$

or, more explicitly,

$$P_i^{\omega_1 + \omega_2}(t) = \frac{1}{4}\left[\chi_{ijk} E_j^{\omega_1} E_k^{\omega_2} e^{j(\omega_1 + \omega_2)t} + \chi_{ikj} E_k^{\omega_2} E_j^{\omega_1} e^{j(\omega_2 + \omega_1)t} + cc\right]$$

In a lossless (instantaneous response) system $\chi_{ikj} = \chi_{ijk}$ and we have

$$P_i^{\omega_1 + \omega_2}(t) = \frac{1}{4} P_i^{\omega_3} e^{j\omega_3 t} + cc = \chi_{ijk} E_j^{\omega_1} E_k^{\omega_2} e^{j(\omega_1 + \omega_2)t} + cc$$

$$P_i^{\omega_3 = \omega_1 + \omega_2} = \chi_{ijk} E_j^{\omega_1} E_k^{\omega_2}$$

In a lossy and consequently dispersive system χ_{ijk} will in general depend on ω_1 and ω_2 and on whether the sum or difference of them is considered. We may thus in general define the nonlinearity not in terms of the instantaneous optical fields but in terms of their complex amplitudes:

$$P_i^{\omega_3 = \omega_1 + \omega_2} = \chi_{ijk}(-\omega_3, \omega_1, \omega_2)\, E_j^{\omega_1} E_k^{\omega_2}$$

$$P_i^{\omega_3 = \omega_1 - \omega_2} = \chi_{ijk}(-\omega_3, \omega_1, -\omega_2)\, E_j^{\omega_1} (E_k^{\omega_2})^*$$

where $\chi_{ijk}(-\omega_3, \omega_1, \omega_2)$ signifies that it is the component of susceptibility relating the ith component of the complex amplitude $P_i^{\omega_3 = \omega_1 + \omega_2}$ to the product $E_j^{\omega_1} E_k^{\omega_2}$. In general

$$\chi_{ijk}(-\omega_3, \omega_1, \omega_2) \neq \chi_{ijk}(-\omega_3, \omega_2, \omega_1)$$

The two coefficients relate to cases where the fields at ω_1 and ω_2 are along different directions; the resultant is not necessarily the same. On the other hand, of course, we have

$$\chi_{ijk}(-\omega_3, \omega_1, \omega_2) = \chi_{ikj}(-\omega_3, \omega_2, \omega_1)$$

Crystals may be divided into two groups depending on whether the crystal structure is invariant with respect to inversion, *i. e.* replacing **r** by **–r**. If it is, crystals are called *centrosymmetric*; otherwise *non-centrosymmetric*.

For transparent lossless crystals the coefficients χ_{ijk} do not depend on the frequencies involved or on whether their sum or difference frequency is generated. In the discussion to follow we shall drop all of their frequency designations.

In the following we shall show that not all crystals will exhibit second harmonic generation. To do this, consider an isotropic crystal. Then χ_{ijk} is a single number χ that is independent of the field polarization. Now reverse the coordinates ($\mathbf{r} \rightarrow -\mathbf{r}$). If we let the electric field and the polarization remain unchanged in direction then their signs must change when we reverse the directions of the axes. That is

$$-P_i^{\omega_1+\omega_2} = \sum_{j,k} \chi_{ijk}^{(2)} (-E_j^{\omega}) (-E_k^{\omega}) = +P_i^{\omega_1+\omega_2}$$

Therefore

$$P_i^{\omega_1+\omega_2} = 0, \ \chi = 0$$

and we conclude that second harmonic generation cannot take place in isotropic media. Further, by the same reasoning, it cannot take place in a crystal whose structure is symmetric about a point. Hence only crystals which lack inversion symmetry exhibit *SHG*. *THG*, however, is possible in crystals which exhibit inversion symmetry.

13.3 Wave propagation in nonlinear media

While the treatment of nonlinear effects in saturable absorbers is often different from that of other materials it is nevertheless instructive to consider the standard formation. We start with Maxwell equations and the constitutive relations for the pairs **E**,**D** and **B**,**H**:

$$\nabla \times \boldsymbol{E} = -\frac{\partial \boldsymbol{B}}{\partial t}, \ \nabla \cdot \boldsymbol{D} = \rho$$

$$\nabla \times \boldsymbol{H} = \boldsymbol{J} + \frac{\partial \boldsymbol{D}}{\partial t}, \ \nabla \cdot \boldsymbol{B} = 0 \qquad 13.1$$

$$\boldsymbol{D} \equiv \varepsilon_0 \boldsymbol{E} + \boldsymbol{P}, \ \boldsymbol{H} \equiv \frac{1}{\mu_0}\boldsymbol{B} + \boldsymbol{M}$$

We separate **P** into its linear and nonlinear parts:

$$\boldsymbol{P} = \boldsymbol{P}_L + \boldsymbol{P}_{NL}$$

$$\boldsymbol{P}_L = \varepsilon_0 \chi^{(1)} \boldsymbol{E} \qquad 13.2$$

$$\boldsymbol{P}_{NL} = \varepsilon_0 [\chi^{(2)} \boldsymbol{EE} + \chi^{(3)} \boldsymbol{EEE} + \chi^{(4)} \boldsymbol{EEEE} + \cdots] \qquad 13.3$$

Using a vector identity and assuming no magnetization present, transversality condition and current-free space, that is, using the following relations

$$\nabla \times \nabla \times \boldsymbol{A} \equiv \nabla(\nabla \cdot \boldsymbol{A}) - \nabla^2 \boldsymbol{A}$$

$$\boldsymbol{M} = 0, \; \boldsymbol{J}_{free} = 0, \; \boldsymbol{E} \perp \boldsymbol{k}$$

we obtain the wave equation

$$\nabla^2 \boldsymbol{E} - \mu_0 \varepsilon \frac{\partial^2 \boldsymbol{E}}{\partial t^2} = \mu_0 \frac{\partial^2 \boldsymbol{P}_{NL}}{\partial t^2} \qquad 13.4$$

Here $\varepsilon = \varepsilon_0 (1 + \chi^{(1)})$ is the tensor linear electric permittivity of the medium. Eq 13.4 shows how electromagnetic waves, which in the linear case passed through one another unaffected by one another, can now be coupled via the nonlinear polarization $\mathbf{P}_{NL}$. This polarization is in general expressed as

$$\boldsymbol{P}(\boldsymbol{r}, t) = \varepsilon_0 \int_{-\infty}^{\infty} \chi^{(1)}(\boldsymbol{r} - \boldsymbol{r}', t - t') \cdot \boldsymbol{E}(\boldsymbol{r}', t') \, d\boldsymbol{r}' \, dt'$$

$$+ \varepsilon_0 \int_{-\infty}^{\infty} \chi^{(2)}(\boldsymbol{r} - \boldsymbol{r}_1, t - t_1; \boldsymbol{r} - \boldsymbol{r}_2, t - t_2) : \boldsymbol{E}(\boldsymbol{r}_1, t_1) \times \boldsymbol{E}(\boldsymbol{r}_2, t_2) \, d\boldsymbol{r}_1 \, d\boldsymbol{r}_2 \, dt_1 \, dt_2$$

$$+ \varepsilon_0 \int_{-\infty}^{\infty} \chi^{(3)}(\boldsymbol{r} - \boldsymbol{r}_1, t - t_1; \boldsymbol{r} - \boldsymbol{r}_2, t - t_2; \boldsymbol{r} - \boldsymbol{r}_3, t - t_3) \,\vdots$$

$$\times \boldsymbol{E}(\boldsymbol{r}_1, t_1) \, \boldsymbol{E}(\boldsymbol{r}_2, t_2) \, \boldsymbol{E}(\boldsymbol{r}_3, t_3) \, d\boldsymbol{r}_1 \, dt_1 \, d\boldsymbol{r}_2 \, dt_2 \, d\boldsymbol{r}_3 \, dt_3 + \cdots \qquad 13.5$$

This expression can be Fourier transformed into **k** and ω space to yield

$$\boldsymbol{P}^{(1)}(\boldsymbol{k}, \omega) = \chi^{(1)}(\boldsymbol{k}, \omega) \cdot \boldsymbol{E}(\boldsymbol{k}, \omega)$$

$$\boldsymbol{P}^{(2)}(\boldsymbol{k}, \omega) = \chi^{(2)}(\boldsymbol{k} = \boldsymbol{k}_i + \boldsymbol{k}_j; \omega = \omega_i + \omega_j) : \times \boldsymbol{E}(\boldsymbol{k}_i, \omega_i) \, \boldsymbol{E}(\boldsymbol{k}_j, \omega_j) \qquad 13.6$$

$$\boldsymbol{P}^{(3)}(\boldsymbol{k}, \omega) = \chi^{(3)}(\boldsymbol{k} = \boldsymbol{k}_i + \boldsymbol{k}_j + \boldsymbol{k}_l; \omega = \omega_i + \omega_j + \omega_l) \,\vdots$$

$$\times \boldsymbol{E}(\boldsymbol{k}_i, \omega_i) \, \boldsymbol{E}(\boldsymbol{k}_j, \omega_j) \, \boldsymbol{E}(\boldsymbol{k}_l, \omega_l)$$

However, in the electric dipole approximation, **P** is a function of the local field only which, when transformed, implies that

$$\chi^{(n)}(\boldsymbol{k}, \omega) \rightarrow \chi^{(n)}(\omega)$$

Thus, at a given frequency ω, the elements of the $\chi^{(n)}$ tensors are constants (complex, in general) for a given "material state". For transparent materials, that is in frequency ranges where absorption is very small, the various orders of χ are usually separated in magnitude by several orders of magnitude. Hence, for a particular set of amplitudes of the interacting electromagnetic waves, one order of the electric susceptibility expansion usually dominates and we need only to consider one corresponding equation for the induced nonlinear polarization.

On the other hand, for strongly absorbing media, the material is changed by the absorption of incident radiation. Since every order of χ is a function of the material state the total changes due to contributions from many terms and the series expansion is less useful. This is the case for saturable absorbing media when the absorption is due to operation near an atomic resonance. Isolating one order of the series expansion of χ is not possible so a full quantum mechanical treatment, including atomic level transitions is required.

13.4 Optical second order harmonic generation (*SHG*)

The first experimental observation of SHG (0.3471 μm) was of a ruby beam (0.694 μm) by Franken and his colleagues. The conversion efficiency in this first experiment was 10^{-8}; since then, however, efficiencies approaching unity have been reported.

Second harmonic generation is a limiting case of the three wave interaction where two of the three fields are of the same frequency, *i. e.* $\omega \equiv \omega_1 = \omega_2$ and $\omega_3 = 2\omega$. Figure 13.1

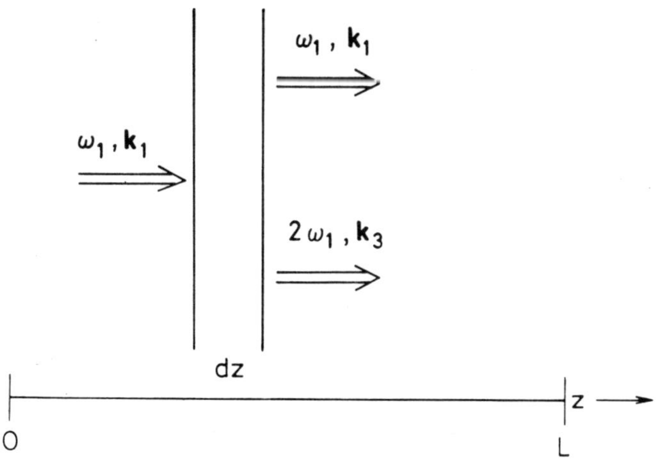

Figure 13.1 Second order harmonic generation

shows a fundamental wave at frequency ω_1 propagating in the *z*-direction deriving a second wave at frequency $2\omega_1$. Let the length of the piece of material be *l*. The strength of the second harmonic signal dE produced within a slab of width *dz* located at *z* is

$$d\mathbf{E}^{(2)}(z) \approx \mathbf{P}^{(2)}(z)\ dz \tag{13.7}$$

$\mathbf{P}^{(2)}$ is the second harmonic polarization induced at frequency ω and is proportional to the square of the electric field **E**.

Thus, corresponding to an angle θ_m, $n_e^{(2\omega)} = n_o^{(\omega)}$, and if the fundamental beam (at ω) is launched along θ_m as an ordinary ray the 2ω-beam will be generated along the same direction as an extraordinary ray. The incident and second harmonic waves are said to be *phase-matched.* The coherence length for second harmonic generation may be made quite large by phase matching. Nonlinear crystals which are of practical interest are the III-V compounds (GaAs, InSb, *etc*), the II-VI compounds (ZnS, CdSe, *etc*), the I-III-VI compounds ($AgGeS_2$, $CuInS_2$, *etc*), the II-IV-V compounds ($CdSiAs_2$, $ZnGeS_2$, *etc*), KDP isomorphs (KH_2PO_4, RbH_2, AsO_4, *etc*), the ferroelectrics ($LiNbO_3$, Ba_2NaNb_5. $LiIO_3$, *etc*), the highly anisotropic crystals (Se, Te, HgS, *etc*), proustite, *etc*. These crystals have large $\chi^{(2)}$. Depending on their regions of transparency some are useful in the visible and others in the infrared.

The second harmonic radiation propagates with wave number k_3 where, in general, $k_3 \neq 2k_1$ because of dispersion. Thus the second harmonic wave generated at plane z_1, having propagated to another plane z_2, is not in phase with the second wave generated at z_2. This results in interference described by the factor

$$\frac{\sin^2 \frac{1}{2}\Delta k\, l}{\left(\frac{1}{2}\Delta k\, l\right)^2}$$

leading to maxima and minima with variation in *l*. Adjacent peaks of this spatial interference pattern corresponding to $\frac{1}{2}(\Delta k)\, l = \left(p + \frac{1}{2}\right)\pi$ are separated by a length l_c, called the *coherence length*, given by

$$l_c = \frac{2\pi}{\Delta k} = \frac{2\pi}{k^{(2\omega)} - 2k^{(\omega)}} \tag{13.8}$$

The coherence length is thus a measure of the greatest crystal length that is useful in producing second harmonic power. For ordinary crystals it may be no more than a few micrometres. This is because the index of refraction *n* normally increases with ω. We have

$$\Delta k = k^{(2\omega)} - 2\,k^{(\omega)} = \frac{2\omega}{c}\left(n^{(2\,\omega)} - n^{(\omega)}\right),\; k^{(\omega)} = \omega\, n^{(\omega)}/c \qquad 13.9$$

The coherence length therefore is given by

$$l_c = \frac{c\,\pi}{\omega\,(n^{(2\,\omega)} - n^{(\omega)})} = \frac{\lambda}{2\,(n^{(2\,\omega)} - n^{(\omega)})} \qquad 13.10$$

where λ is the free space wavelength of the fundamental beam. Considering typical values of $\lambda = 1$ μm, $[n^{(2\omega)} - n^{(\omega)}] = 10^{-2}$ we get l_c=100 μm. If it were to increase to, say, 2 cm second harmonic power would go up by a factor of 10^4.

The technique that is used widely to satisfy the phase-matching requirement, *viz* $\Delta k = 0$, takes advantage of the natural birefringence of anisotropic crystals, using the relation $k^{(\omega)} = \omega\,\sqrt{(\mu\varepsilon_0)}\,n^{(\omega)}$ in $k^{(2\,\omega)} = 2\,k^{(\omega)}$ so that we have

$$n^{(2\omega)} = n^{(\omega)} \qquad 13.11$$

Thus the indices of refraction of the fundamental and second harmonic frequencies must be equal. In normally dispersive materials the index of the ordinary wave or the extraordinary wave along a given direction increases with ω and eq 13.11 cannot be satisfied if both ω and 2ω beams are of the same type, *i. e*. both ordinary or both extraordinary. However,

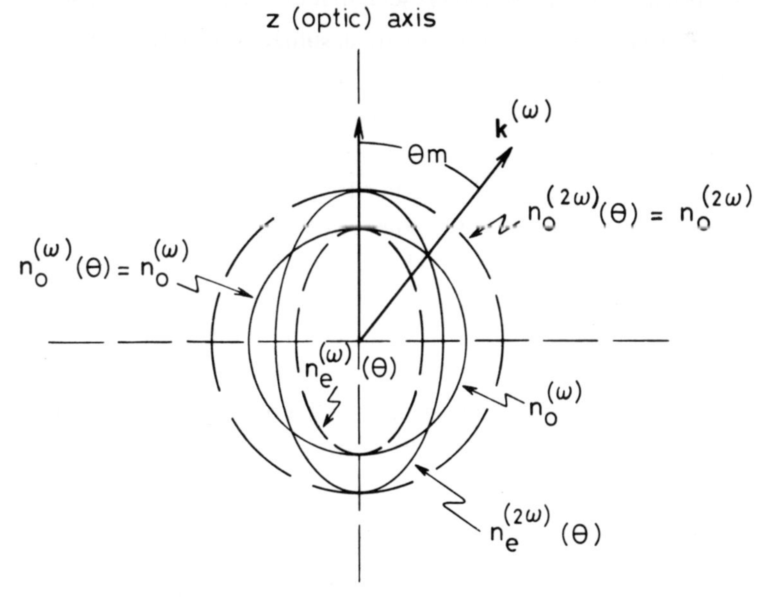

Figure 13.2 Ray surface for second harmonic generation

the equation can be satisfied if they are dissimilar. The example of KDP may be cited. The incident wave may be introduced into the crystal as an ordinary one. The 2ω wave will be an extraordinary one. The ray surface – sphere – for an ordinary ray with frequency ω crosses that – ellipsoid – for an extraordinary ray with frequency 2ω as shown in Figure 13.2 (p. 203). Therefore, if the incident wave is made to propagate through the crystal precisely at the angle θ_m shown in the figure, $\Delta n = [n^{(2\omega)} - n^{(\omega)}]$ will be nearly zero.

13.5 Optical mixing

In the analysis of SHG we considered the nonlinear polarization that results from the interaction of two fields having identical frequencies. If they have different frequencies the nonlinear polarization and, therefore, the light emitted contains sum and difference frequencies. This process is known as *parametric amplification.* The basic configuration involves an input "signal" at ω_1 that is incident on a nonlinear crystal together with an intense pump wave at $\omega_3 > \omega_1$. The amplification of the former wave is accompanied by the generation of an "idler" wave at $\omega_2 = \omega_3 - \omega_1$.

As with SHG, here too, phase-matching is extremely important; indeed the requirement is more stringent because of the number of frequencies involved. In the case of parametric amplification three rather than two waves must be phase-matched. Thus for a pump wave at ω_3 to provide – via interaction in a nonlinear crystal – simultaneous amplification for optical waves at ω_1 and ω_2 such that $\omega_3 = \omega_1 \pm \omega_2$ we must also satisfy the condition $k_1 \pm k_2 = k_3$ (where the three waves are assumed to be collinear). If the nonlinear crystal is placed within an optical resonator (*see* Chapter 12) the power from the parametric amplifier can be increased. An optical resonator provides resonances for the signal or idler wave (or both) so the parametric gain will, at some threshold pumping irradiance, cause simultaneous oscillation at both signal and idler frequencies. The threshold for this oscillation, similar to a laser, corresponds to the point at which the parametric gain just balances the losses of the signal and idler waves. The process is therefore known as *parametric oscillations*. The parametric oscillator can be tuned either by rotating the crystal so that different frequencies are phase-matched or by changing the temperature of the crystal so that its properties change slightly. Thus the practical device-importance of a parametric oscillator is its ability to convert the power output of a pump laser to a coherent output at the signal and idler frequencies -which can be tuned continuously over wide ranges.

13.6 Frequency upconversion

Parametric interactions in a crystal can be used to convert a signal from a low frequency to a higher one. Two collinear beams with frequencies ω_1 and ω_2 are shone on a nonlinear crystal. If the crystal is phase-matched for these frequencies and the sum frequency ω_3 a

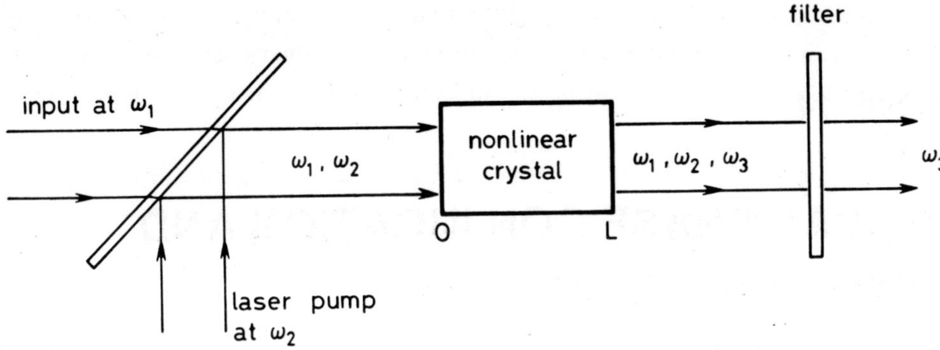

Figure 13.3 Experimental arrangement for frequency upconversion

third wave whose frequency is ω_3 will be generated. The first two waves may be filtered out if necessary. Figure 13.3 shows an experimental arrangement. The ω_1 and ω_2 beams are combined in a partially transmissive mirror so that they traverse together, in parallelism, the path length l in a nonlinear crystal. One of the main applications of frequency upconversion is to detect radiation from IR and FIR lasers, at which frequencies fast or sensitive detectors are often not available. The radiation from the laser is converted to visible or near IR radiation that is then easily detected.

13.7 Optical mixing in vapour

In a vapour system, because of inversion geometry, $\chi^{(2)}$ vanishes in the electric dipole approximation. One would expect the next higher order nonlinearity $\chi^{(3)}$ to be small. However, when optical frequencies are close to resonances it increases – often as strongly as $\chi^{(2)}$ does in crystals – due to resonance enhancement. Over the past few years there have been many reports on third order optical mixing in atomic vapours. Atomic systems now appear to be the type of media which can be used to generate uv or even x-ray radiation by nonlinear optical means. In a typical experiment an N_2-pumped dye laser is used to obtain tunable vacuum uv radiation. The N_2 laser is used to pump two dye lasers simultaneously, one at frequency, say, ω_1 and the other at, say, ω_2. With both frequencies variable, tunable uv output at $2\omega_1+\omega_2$ from Sr vapour has been observed over a spectral range of several thousand cm^{-1}.

Bibliography

SHEN Y R 1976 Rev Mod Phys **48** 1

YARIV A 1985 *Optical Electronics* 3rd ed (Tokyo: Holt Saunder)

Chapter 14

OPTICAL PHASE CONJUGATION AND FOUR-WAVE MIXING

K C RUSTAGI

14.1 Introduction

One of the most remarkable and revolutionary experiments in nonlinear optics was undoubtedly the first observation of optical phase conjugation and its distortion-correction or wavefront-restoring properties. It was done by Zel'dovich *et al* (1972) at the Lebedev Institute. Trying to understand their earlier observation that the back-scattered wave in stimulated Brillouin scattering *(SBS)* propagated in the same solid angle as the input wave, they looked at the correlation between the input and output wavefronts. The experimental setup used for this is shown schematically in Figure 14.1 (*see* next page). The beam splitter BS serves to record the input and output beam profiles separately. A is a ground glass plate used to distort or aberrate the beams passing through it. The 1m long cell C contained methane gas at 125 atmospheric pressure. They observed that the reflected SBS wave had, after passing through the distorting plate A, almost the same spatial quality as the input laser beam before it was distorted by A. Thus the reflected SBS wavefront was such that it produced an undistorted wavefront on passing through the aberrating glass A. This would happen if the output SBS beam had a wavefront that was a replica of that of the pump beam but travelling in the opposite direction. As we shall see, this is equivalent to saying that the output wave amplitude is proportional to the complex conjugate of the input wave amplitude. The process of generating such oppositely travelling replica wavefronts is called *optical phase conjugation.*

This amazing discovery of a method of wavefront correction did not receive any great attention immediately although Nosach *et al* (1972), from the same group, had demonstrated that the method could be used to generate good quality amplified beams from relatively poor quality (inhomogeneous) laser amplifiers. SBS is an efficient process and has been studied a great deal as a method of optical phase conjugation. We note that the phase conjugate beam has a slightly shifted frequency compared to that of the incident one.

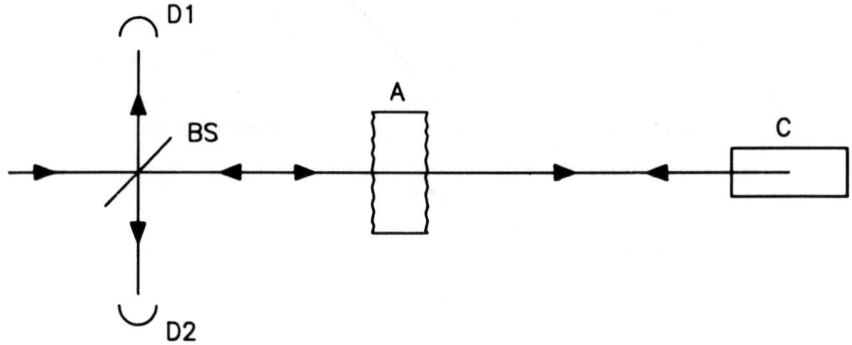

Figure 14.1 Schematic representation of the experimental setup used in the first observation of wavefront recovery by phase conjugation. The cell contained a hollow glass light pipe to ensure long interaction length (adapted from Ref 1).

Several reviews of this and other methods of nonlinear optical phase conjugation *(OPC)* are available (collected together in *References* for convenience under the earliest referred to, *viz* Shen 1984).

A spurt in OPC studies came with the work of Hellwarth (1977) who showed that the phase-conjugated beam could be obtained by DFWM also. Since then this process has been investigated vigorously as a means of OPC as well as for many other applications. In the following we shall concentrate on DFWM as a method of obtaining phase-conjugated laser beams.

To understand the phenomenon of DFWM consider two counter-propagating laser beams, called the forward (f) and backward (b) pump waves respectively, and a third wave, called the probe wave (p), all having the same frequency ω intersecting in a nonlinear medium as shown in Figure 14.2 (*see* next page). We look for a fourth wave, also at ω, generated as a result of this interaction. All of the four waves — three input and one output — have the same frequency (hence the nomenclature *degenerate* four-wave mixing) and are distinguished from each other by their propagation vectors $\mathbf{k}_f$, $\mathbf{k}_b$, $\mathbf{k}_p$ respectively for the input waves and $\mathbf{k}_s$ for the output or signal wave. As in the standard theory of nonlinear optical wave propagation, let the electric fields $\mathbf{E}_i$ in the input waves (i = f, b, p) be written as

$$\boldsymbol{E}_i\ (\boldsymbol{r}, t) = \boldsymbol{E}_i\, e^{-j(\omega t - \boldsymbol{k}_i \cdot \boldsymbol{r})} + complex\, conjugate\ (c\ c) \qquad 14.1$$

The total self-consistent field consists of all of the input waves and all of the generated waves. Its Fourier component ω obeys the wave equation

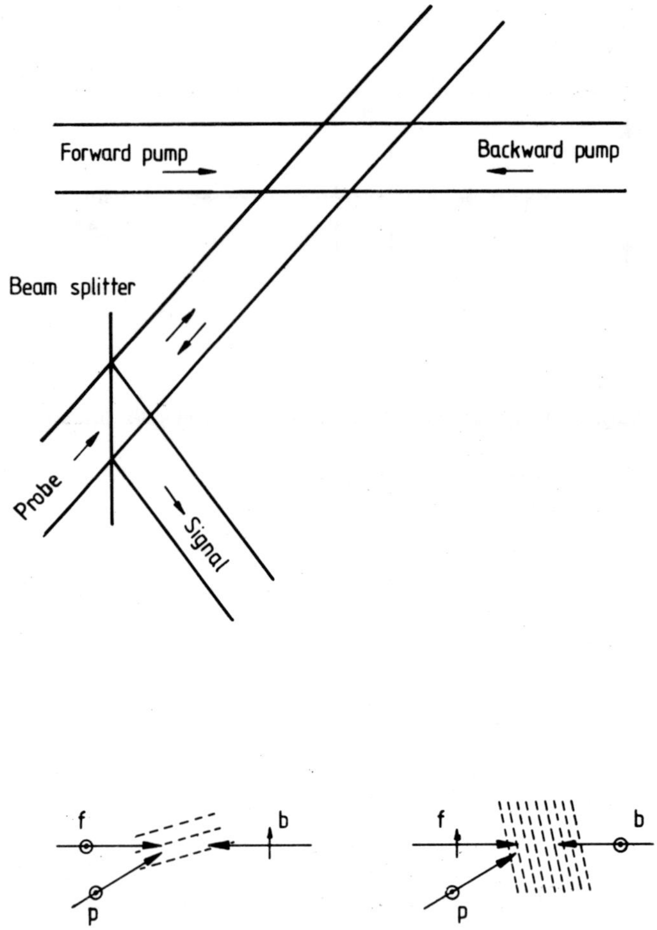

Figure 14.2 Standard geometry for degenerate four-wave mixing experiments with polarization configurations for exciting short period and long-period transient gratings

$$\nabla \times \nabla \times \boldsymbol{E}(\omega) - (\omega^2 / c^2)\, \varepsilon(\omega)\, \boldsymbol{E}(\omega) = (4\pi \omega^2 / c^2)\, \boldsymbol{P}^{NLS}(\omega) \qquad 14.2$$

where $\varepsilon(\omega)$ is the appropriate dielectric function at frequency ω. The nonlinear source polarization $\mathbf{P}^{NLS}(\omega)$ is given by

$$\boldsymbol{P}^{NLS}(\omega) = \chi^{(3)}(-\omega, \omega, -\omega, \omega) : \boldsymbol{E}(\omega)\, \boldsymbol{E}^*(\omega)\, \boldsymbol{E}(\omega) \qquad 14.3$$

As a first approximation, we may neglect the modification of incident wave amplitudes and write them as

$$\boldsymbol{E}(\omega) = \sum_{i = f, b, p} A_i\, e^{j \boldsymbol{k}_i \cdot \boldsymbol{r}} \qquad 14.4$$

Then $\mathbf{P}^{NLS}$ contains terms with several different propagation vectors. One of these is the term

$$P^{NLS}(\omega) = \chi^{(3)} : A_f A_p^* A_b e^{j(k_f - k_p + k_b)\cdot r} \qquad 14.5$$

The polarization wave propagates with a wavevector $\mathbf{k}_s = -\mathbf{k}_p$ since $\mathbf{k}_f + \mathbf{k}_b = 0$ and thus creates a signal wave which travels in a direction exactly opposite to that of the probe wave. In terms of photons, the elementary process involved is the scattering of two pump photons, also of zero total momentum, one to amplify the probe beam and the other to create the new beam. Designating the signal by the subscript s, the relevant energy and momentum conservation equations are

$$\hbar(\omega_f + \omega_b) = \hbar(\omega_s + \omega_p)$$

$$\hbar(k_f + k_b) = \hbar(k_s + k_p)$$

It is of some interest to note here that the phase matching or momentum conservation condition can also be satisfied in a nonplanar geometry as shown in Figure 14.3 where

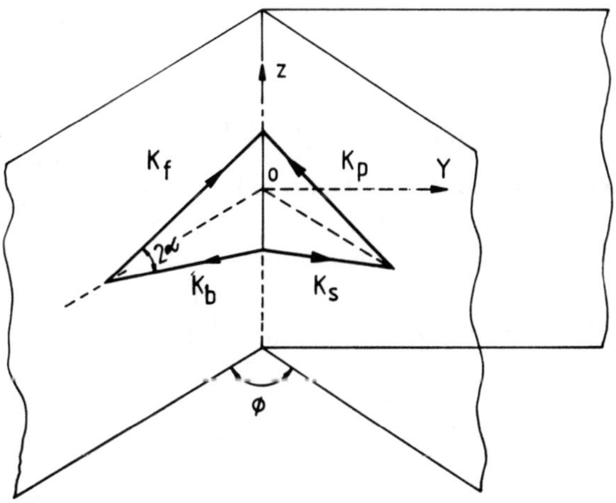

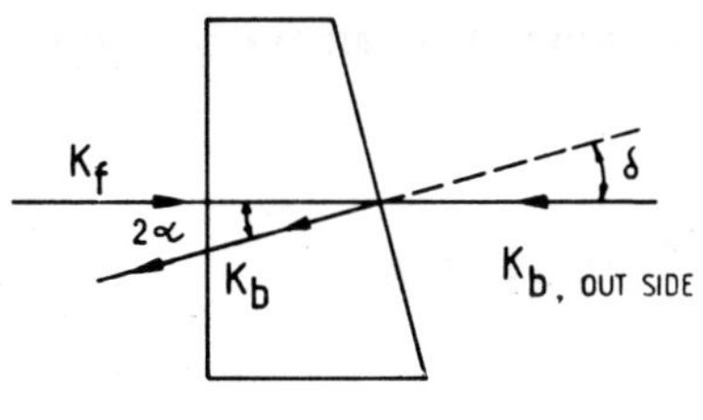

Figure 14.3 Noncoplanar phase-matching for DFWM. The inset shows misalignment of pumps caused by a wedge.

$\mathbf{k}_f + \mathbf{k}_b = \mathbf{k}_p + \mathbf{k}_s \neq 0$. Sometimes this can happen unintentionally either because the two pump beams are not aligned sufficiently accurately or because the sample has a small wedge between the two supposedly parallel faces. The pump beams which are aligned outside the sample get misaligned in the sample as shown in the insert of Figure 14.3. In such a situation the signal beam does not return along the probe beam exactly. The coplanar equivalent of this phase matching configuration is not the usual DFWM configuration but the socalled cross-modulation configuration ($\mathbf{k}_p = \mathbf{k}_b \neq \mathbf{k}_f = \mathbf{k}_s$). The probe beam is incident collinearly with one of the pump beams and the signal emerges along the other pump beam. To detect the process the probe beam should be modulated and the modulated signal wave may be detected after removing the constant pump wave.

Returning to eq 14.5 we can see that if the probe field carries a slowly varying phase factor exp $[-j\,\phi\,(\mathbf{r})]$, where $\mathbf{r}$ is the position vector in the plane transverse to the direction of propagation, the nonlinear source polarization and hence the generated wave will be proportional to that phase factor.

More specifically, taking z as the direction of the probe wave traversing a nonlinear medium of extension $0 \leq z \leq L$, uniformly illuminated by two counter-propagating pump beams, it can be shown that the amplitude of the signal wave generated is given (Caro and Gower 1982, Gower 1985) by

$$A_s(z) = \frac{2jQA_p^*(0)\,\sin[H(z-L)/2]\,e^{-\alpha z/2}}{H\cos(HL/2)+\alpha\sin(HL/2)}, \quad 0 \leq z \leq L \qquad 14.6$$

where α is the absorption coefficient at ω, Q the nonlinear coupling coefficient given by

$$Q = -\frac{4\pi\omega}{2cn}\,3\chi^{(3)}\,|E_f(0)\;E_b(L)|\;e^{-\alpha L/2\cos\theta} \qquad 14.7$$

$$H^2 = 4|Q|^2 - \alpha^2 \qquad 14.8$$

Here θ is the angle between the forward pump and the probe beams. All incident waves have been taken as polarized perpendicular to the plane containing the three incident beams and it has been assumed that all the amplitudes and phase $\phi\,(x, y)$ are slowly varying functions of z.

14.2 Wavefront Recovery

It is clear from eq 14.6 that in the $z = 0$ plane the signal wave amplitude is proportional to the complex conjugate of the probe wave amplitude in that plane. From this it can be shown (Yariv 1985) that in the scalar wave theory the signal wave amplitude is the complex conjugate of the probe wave amplitude throughout the half-space $z \leq 0$. Thus for a probe wave described by

$$E_p(\mathbf{r}, t) = E_p\, e^{j\varphi(x,y)}\, e^{-j(\omega t - kz)} + cc \qquad 14.9$$

the signal wave is given by

$$E_s(\mathbf{r}, t) = E_s\, e^{-j\varphi(x,y)}\, e^{-j(\omega t + kz)} + cc \qquad 14.10a$$

which can also be written as

$$E_s(\mathbf{r}, t) = E_s\, e^{j\varphi(x,y)}\, e^{j(\omega t + kz)} + cc \qquad 14.10b$$

Thus, apart from a multiplicative factor, the signal wave is a time-reversed replica of the incident probe wave, *i. e.* at every given $z \leq 0$ the incident wave and the conjugate wave have the same wavefront although the two directions of propagation are mutually opposite. Even when this signal wave passes through a linear phase-distorting medium, *e. g.* an aberrating glass plate, this property of time reversal is maintained due to the time reversal invariance of the wave equation so long as scattering is negligible. Hence the final output beam has the same wavefront as the incident one.

In terms of geometrical optics one can say that a phase conjugate mirror reflects each and every incident ray back on itself. Thus converging rays diverge on reflection by a phase

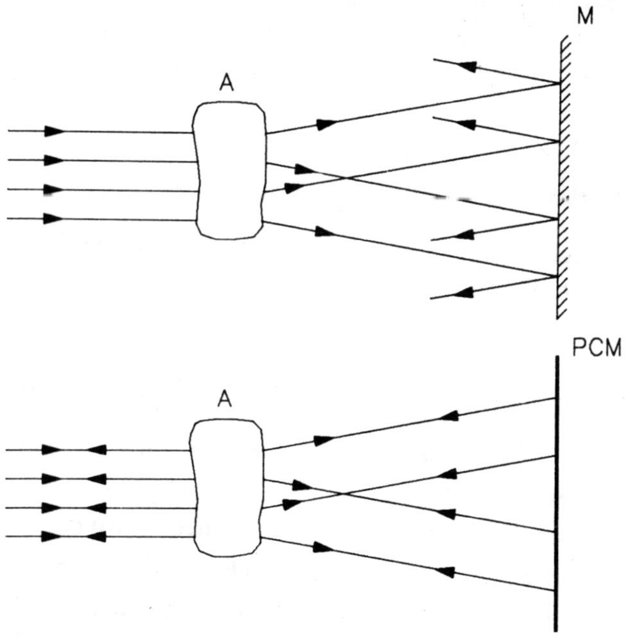

Figure 14.4 Geometrical optic representation of ordinary and phase conjugate mirrors

conjugate mirror and rays radiating from a point source would always focus back to the point no matter what linear and transparent optical elements are put between the phase conjugate mirror (*PCM*) and the source. That this is so is easily seen in the DFWM scheme where we note that because of the phase-matching condition the generated beam travels back along the probe beam irrespective of the angle of incidence of the probe beam. Figure 14.4 (p. 211) illustrates how this ability to send every ray back on itself helps undo or heal the distortion caused by an inhomogeneous transparent medium. Parallel incident rays are refracted into different directions by an aberrator A, an irregular-shaped or inhomogeneous transparent object. On reflection by a normal mirror M the directions of the various reflected rays are determined by the corresponding angles of incidence at the mirror. Only some of these rays will pass through A on the return path and even these will not emerge parallel to one another. On the other hand, a phase conjugate mirror (PCM in Figure 14.4) reflects each ray back on itself. The reflected rays thus retrace their paths through the aberrator and emerge parallel to one another from the aberrator. We should, however, note that we have assumed that the aberrator preserves the polarization purity of the beam. Since usually an inhomogeneous medium would cause polarization scrambling, considerable effort has been devoted in the last few years (Tomita *et al* 1989) to evolve OPC schemes which reverse the wavefront for an arbitrarily polarized probe beam. We have also neglected back scattering by the aberrator (Wolf *et al* 1983).

This brief discussion already shows that a phase conjugate mirror is a fundamentally new optical element and opens up revolutionary possibilities. We discuss some of them in the following. Of course, some properties of a phase conjugate mirror such as reflectivity, polarization and spectral response depend on the nonlinear optical material used for the process.

14.3 DFWM as real time holography

Degenerate four-wave mixing is sometimes called *real time holography*. To see this connection let us note that for isotropic media the nonlinear source polarization can be written (Giuliano 1981) as

$$\boldsymbol{P}^{NLS}(\omega) = A(\theta)\,(\boldsymbol{E}_f \cdot \boldsymbol{E}_p^*)\,\boldsymbol{E}_b\; A(\pi - \theta)(\boldsymbol{E}_b \cdot \boldsymbol{E}_p^*)\,\boldsymbol{E}_f + B\,(\boldsymbol{E}_f \cdot \boldsymbol{E}_b)\,\boldsymbol{E}_p^* \qquad 14.11$$

where θ is the angle between the f and p beams. The first two terms here have simple holographic analogues as can be seen from the following consideration. If we look at the irradiance pattern created by the simultaneous presence of two beams f and p we will find a stationary interference pattern depicted by $\mathrm{Re}(\mathbf{E}_f \cdot \mathbf{E}_p^*)$ superposed on a constant background proportional to $|\mathbf{E}_f|^2 + |\mathbf{E}_p|^2$. The irradiance-dependence of the refractive index of the medium converts this irradiance pattern into a refractive index pattern or a volume hologram (*vide* §§ 5.14-15, 5.22-23). If f and p are both plane waves the interference pattern is a grating with period $2\pi \mid \mathbf{k}_f - \mathbf{k}_p \mid$. The third beam b can now diffract from this volume phase grating and following the Bragg condition create a beam directed opposite to the probe beam. Similarly a p–b grating diffracts the forward pump beam. The connection between DFWM and holography is now quite obvious. The object wave in holography is now replaced by the probe beam, the reference wave by one of the two pump beams and the readout beam by the other pump beam. The difference is that recording is not permanent and reading may be done simultaneously. The second difference, of course, is that in DFWM one does not require to develop the hologram. Most of the tricks of image manipulation used in holography can be adopted and information processed in real time. Clearly, for the volume hologram to exist any modification of the complex refractive index would do; we thus have a phase hologram when only the real part of the refractive index depends on the irradiance and an absorption hologram when the absorption is irradiance-dependent.

The third term in eq 14.11 is not interpretable in terms of a temporally stationary grating unlike the first two terms. It occurs even if the probe beam is polarized perpendicular to both the pump beams and is due to coherent interaction among the three beams. When the frequency ω as well as 2ω is away from all the resonances of the medium the three terms in eq 14.11 are expected to be of comparable magnitude. However, when ω is close to a resonance frequency of the medium the grating contributions dominate. In the case of resonant excitation real transfer of energy occurs from the laser to the medium. The irradiance pattern created by the interference of the two "writing" beams is transformed into an "excitation pattern". The excited medium has a different complex dielectric function compared to the unexcited state. The change in the real part corresponds to the phase-modulating volume gratings and that in the imaginary part to absorption-modulating gratings.

In this situation the time-dependence of the DFWM process is characterized by at least three important time constants. The response time or rise time is the time constant characterizing the establishment of the grating. The second time constant – the recovery time– characterizes the fall time of the nonlinear polarization in the medium. In a non-resonant situation the two are expected to be the same and are determined by the frequency of the dominant virtual excitation. However, when real excitation occurs in the medium these two times can be very different. Moreover, a third time constant now becomes important if the medium is homogeneous and the excitation can move through the sample. The ability of the excitation to move implies that the spatial variation of the excitation can vanish even before de-excitation of the medium. The corresponding characteristic time is called the *diffusion time* or the *grating washout time*.

For illustration let us consider the case of a semiconductor excited by a laser with photon energy $\hbar\omega \geq E_g$, the band gap. The excitation in this case is the electron-hole pair. An electron-hole pair may diffuse before it can recombine. The diffusion of the excited state implies a washing out of a sinusoidal grating in a time of the order of $\tau_D = \Lambda^2 / 4\pi^2 D_a$, where D_a is the ambipolar diffusion coefficient and Λ the period of the grating (Jain and Klein 1982). It is important to note at this stage that the three terms in eq 14.11 can be separated by choosing appropriate polarizations of the three input beams. If, for example, the forward pump and probe beams are polarized parallel to each other and the backward pump is polarized perpendicular to both, only the f-p grating is formed and the signal beam, being due to the scattering of the backward pump from this grating, is polarized perpendicular to the probe beam. Using pulsed lasers the two grating terms can also be distinguished from each other by delaying one of the two pump pulses with respect to the probe.

In a typical experimental setup for efficient phase conjugation the probe beam makes a small angle with respect to one of the pump beams, say the forward pump beam, so that the interaction volume is reasonably large. Then the periodicities of the two gratings (b-p and f-p) are somewhat different. For $\theta << 1$, a quick estimate gives the periods of the two gratings as $\lambda/2n$ and λ/n, 17 where λ is the wavelength of the laser radiation in the medium and θ the angle between the probe beam and the forward pump beam also in the medium. In this type of situation the f–p grating is also the slower to disappear if the diffusion time τ_D is much smaller than the recombination time τ_R. Consequently the response of the f-p grating is much greater than that of the b-p grating.

It is pertinent to note that from this viewpoint of transient holograms DFWM has been studied in semiconductors (Woerdman and Bolger 1969, Woerdman 1970) and saturable absorbers (Eichler *et al* 1986) even before the work of Hellwarth (1977). Indeed Woerdman (1970) had also noted that the disappearance of the gratings is primarily due to ambipolar diffusion of electron-hole pairs. The method has often been used in the investigation of electron-hole diffusion in semiconductors (Eichler *et al* 1986, Weber *et al* 1988). A very interesting perspective on laser-induced gratings in a variety of situations is given by Eichler *et al* (1986).

14.4 Semiconductor-doped glasses

Since optical nonlinearities due to electron-hole pairs generated by a laser near the absorption edge are known to be large there has been some concern about the deleterious effects of carrier diffusion in semiconductors. In the steady state regime (laser pulse duration > τ_D, τ_R) the efficiency of DFWM is often restricted by τ_D. Diffusion also determines the minimum pixel size in optical logic applications. More importantly, it introduces an extraneous distinction between the f-p and b-p gratings so that any optical logic scheme or any other application exploiting polarization-dependence of the DFWM processes becomes unworkable. To overcome carrier diffusion Jain and Lind (1983) used semiconductor particles doped in borosilicate glasses as nonlinear materials.

They used the second harmonic of a Q-switched Nd:YAG laser as a source for experiments on the commercial filter 3483 or C3-68 of Corning glass. Surprisingly, the nonlinearity observed was found to be faster than the laser pulse duration. Even at the lowest fluence used by them the DFWM signal showed some saturation effect. Polarization-dependence of the signal showed that, as anticipated, the two gratings, *viz* the large-period grating formed by the forward pump and probe beams and the small-period grating formed by the backward pump and probe beams, contributed almost equally. This established that the electron-hole pairs created by the light beam did not diffuse. However, when both gratings were present, the signal observed was less than that expected if the two gratings added in phase. This was subsequently explained as being due to self phase modulation of the input beams (Hache *et al* 1987).

Soon after the work of Jain and Lind, Rustagi and Flytzanis (1984) extended the well known Maxwell-Garnett theory (1904, 1906) – appropriate for linear optical properties of such composite materials – to the nonlinear regime. They noted the possibility of resonant enhancement of such nonlinearities and also a new mechanism of nonlinearity *viz* the irradiance-dependence of field penetration. Since then many experimental efforts have been devoted to understanding and exploiting the optical nonlinearity in commercial glass filters containing CdS_xSe_{1-x} crystallites.

Following the theoretical analysis by Rustagi and Flytzanis, Roussignol, Ricard, Rustagi and Flytzanis (1985), henceforth referred to as *RRRF*, repeated the experiments of Jain and Lind with some important modifications. First, they used the second harmonic of a mode-locked and Q-switched Nd:Yag laser with pulse width of 28 ps so that the anticipated fast decay of the DFWM signal could be measured accurately. Second, a spatial filter was used for the DFWM-reflected beam before detection and a much higher signal-to-noise ratio achieved. Finally, coloured glass samples from two different manufacturers were used which showed very different carrier lifetimes. This was confirmed by time-resolve fluorescence measurements. This important observation by RRRF has resulted in many efforts by various techniques generating a variety of conflicting claims. RRRF found a rather slow decay of DFWM in the Schott glass OG 530 (decay time constant ~ ns) while the Corning C3-68 glass showed a fast component with decay time ~ 0.5 ns and a slower one with decay time ~ 11 ns. It is important to mention that both Jain and Lind and RRRF used filters with absorption edge close to the laser frequency so that the carriers generated has negligible excess energy. When the carriers are excited well above the gap they can rapidly relax back towards the band edge through the emission of LO phonons. Thus Nuss *et al* (1986), in their pump-probe measurements, found that the absorption recovery time is much shorter when the excess energy is large. Cotter (1986), on the other hand, found that DFWM signal decay also depends upon irradiance. This has been attributed to a photochemical change or "darkening" and to the Auger recombination process which becomes efficient at higher carrier densities. Both these explanations seem somewhat unsatisfactory.The physical process involved in photochemical darkening is not yet clearly established although it has been speculated that a structural change is involved (Takagahara 1987). Second, it seems difficult to justify the Auger recombination process in a microcrystal in the same way as in the bulk. For a particle of diameter 10 nm or smaller an electron density of 10^{18} cm^{-3} corresponds to less than 1 electron per crystallite. Far from being quasi-continuous, the energy levels in such a particle should exhibit noticeable quantum size effects. We note that a broad distribution in particle size can give sufficient inhomogeneous broadening so that the peaks in absorption are masked effectively; but that does not imply that the carrier dynamics in the particle is also bulk-like since it involves only one particle. More recently Horan and Blau (1990) have reported a systematic study of the darkening process in specially prepared glasses with particle sizes in the range of 2.5 ± 0.3 nm. These materials continue to attract attention. For more details reference may be made to the various review and research papers grouped, for convenience, in *References* under Horan and Blau (*loc cit*).

14.5 Imaging applications

The fact that a PCM reflects every ray on to itself endows it with the capacity of image projection with large numerical aperture. Detailed studies (Levenson and Chiang 1982, Levenson 1983 and references therein) have demonstrated the feasibility of developing a high resolution, large-numerical-aperture optical lithography setup based on a PCM. In the early experiments on image projection a low power continuous wave Argon ion laser was used with photorefractive $BaTiO_3$ as the nonlinear medium. The exposure times required were several hours. Later experiments were performed with the third harmonic Q-switched Nd:YAG laser and Rhodamine 6G solution in acetone was used as the nonlinear medium (Levenson 1983). The mechanism of nonlinearity in this case involves the temperature-dependence of the refractive index. Briefly, the pump-probe interference pattern is converted by absorption and subsequent thermalization into a temperature grating. The temperature-dependence of the refractive index converts that into a phase-modulating grating. The reflectivity of this kind of phase conjugate mirror has been demonstrated to be in excess of 300%. It should be noted that uv lasers are the most effective for photolithography application because of the rich variety of photochemical reactions in the uv and because the shorter the wavelength the higher the resolution achievable.

Another kind of imaging application of DFWM is by virtue of directional output that the process shares with the well known coherent antistokes Raman scattering (*CARS*) which has been used extensively for spatially resolved spectroscopic studies of highly luminescent objects such as flames and plasmas. For example, Ewart *et al* (1986) have used DFWM to image atomic concentration in plane cross-sections of a flame. The two dye laser pumps are focussed cylindrically so that 90% of the pump energy is contained in a thin sheet (~ mm × 0.5 mm). The probe beam illuminates a circular cross-section. The DFWM signal images the distribution of the sodium atoms in the flame with a resolution of up to 100 μm.This application of DFWM spectroscopy is not yet fully explored and much progress can be expected (Rakestraw *et al* 1990). The method is complementary to CARS since in CARS vibrational excitations in molecules are more conveniently probed while in DFWM electronic excitations in atoms or small molecules are more conveniently probed.

14.6 DFWM in optical computing

There are several interesting applications of DFWM in optical computing. As already noted, DFWM can be viewed as real time holography. So it allows holographic optical image processing to be done in real time. This had been discussed in several reviews (Fisher, *Ed*, 1982, Pepper 1985, Yariv 1985). Second, DFWM can be exploited (Yao Li *et*

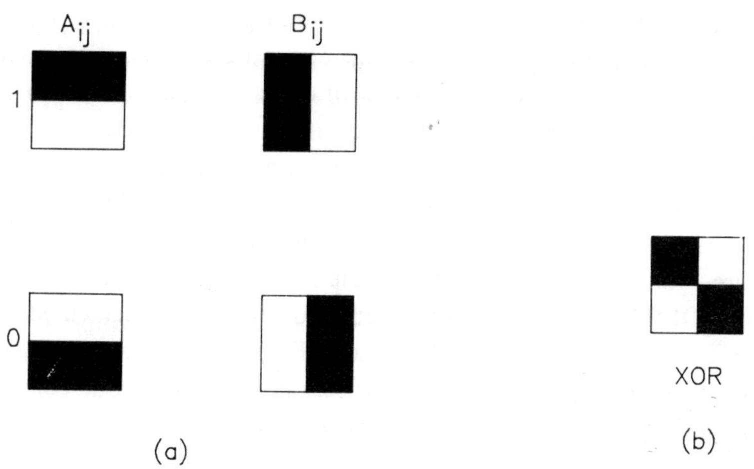

Figure 14.5 Patterns used for binary optical logic by geometric shadowcasting method (adapted from Ref 39)

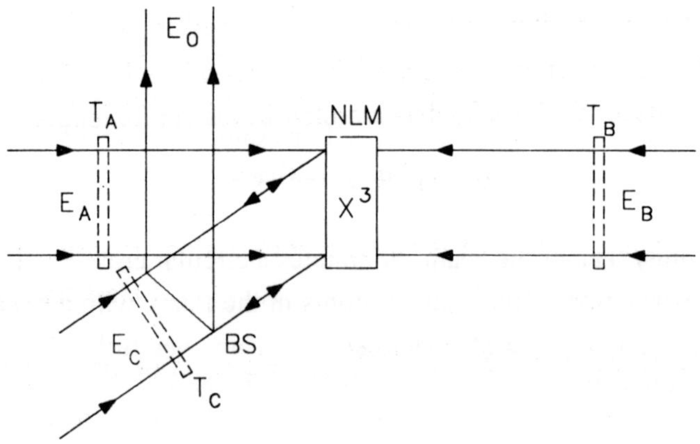

Figure 14.6 Optical implementation of shadow-casting logic using DFWM. Transparencies TA,TB and TC carry the patterns corresponding to inputs A and B and the decoding mask respectively (adapted from Ref 37)

al 1988) in implementing the geometrical optical shadow-casting methods (Yatagai 1986, Ichioka and Tanida 1989) for digital optical logic. In this method the logical inputs A and B are in the form of coded, geometrical patterns. For example, A is represented by a horizontal binary pattern and B by a vertical binary pattern (Figure 14.5a, p. 217). The logic function to be implemented is represented by a 4-cell decoding pattern such as that shown in Figure 14.5b for XOR function. Decoding patterns for other logic functions are given by Yatagai (1986). In the output, logical 1 is the situation when light passes through any of the four cells of the decoding mask superposed with the input masks A and B. A completely dark output field or zero transmission is taken as the logical zero. If this logic is implemented by mechanically superposing the three transparencies — the two input patterns and one decoding pattern — the advantage of parallelism of optical computing is retained but the processing speed, *i e* the number of operations performed per second, is obviously very slow. This can be improved greatly if the transparencies are replaced by electrically addressed or optically addressed spatial light modulators (Neff *et al* 1990). The same process can be performed through DFWM also since the DFWM output is proportional to three inputs which can now be coded into the two pump beams and the probe beam as shown in Figure 14.6. The signal beam carries the logical output information. The main advantage of using DFWM is that several logic functions can be implemented simultaneously by using different probe beams as illustrated in Figure 14.7 (*see* next page) for two functions. Further, since DFWM is a coherent process the method can be combined with the usual coherent analogue techniques (Eichmann *et al* 1987).

Yet another promising optical computing application of OPC is in the realization of optical associative memories (Owechko 1989). In present day digital electronic computers

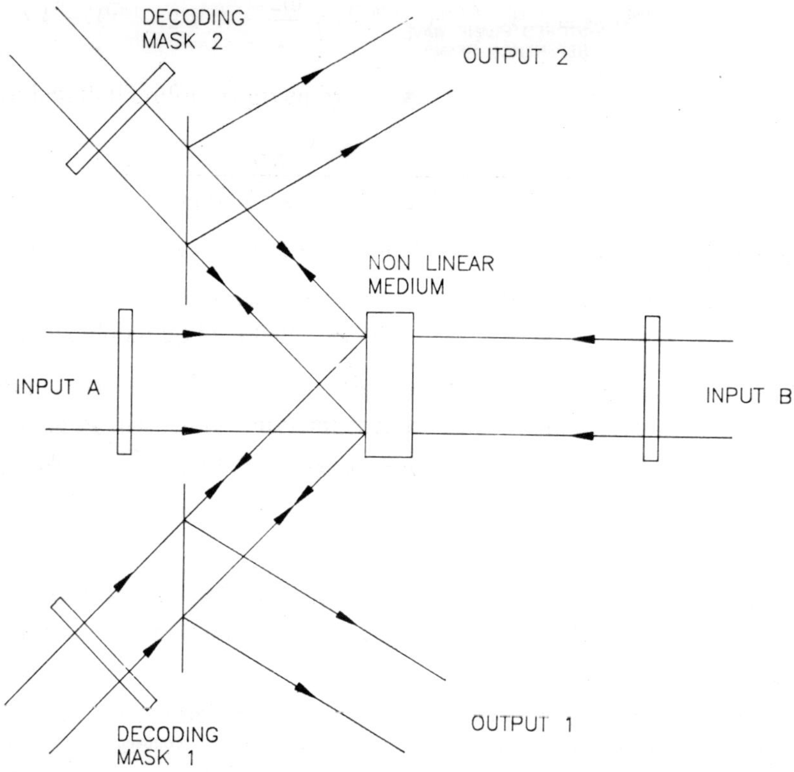

Figure 14.7 A schematic arrangement showing the use of simultaneous multiple (2) function processing

the information is stored in random access memories (*RAM*) in which each memory register is identified by a number. To read or recall the information stored in any given register is simple in such an arrangement. However, if we want to find full information about an item about which we possess part information we have to scan each memory register one by one to check by comparing if the item stored there matches the part information. In contrast, the human brain can recall or recognize people by their faces or their dress or the way they walk; each piece of information is obviously a small part of the information we have about a person. An information storage system that permits direct recall or reconstruction of the complete information about an item from the knowledge of one part of that information is called an *associative* or *content addressable* memory. In one particular realization of such a memory one can recall the complete image of an object when any part of that image is used as input (Owechko 1989). Volume holograms of several images are first recorded sequentially in the same hologram using a distinct reference beam for each object wave-front. The reference beams are all plane waves of the same frequency but with different directions of incidence. Let $\{a_i\}$ be the object waves recorded with $\{b_i\}$ denoting the plane

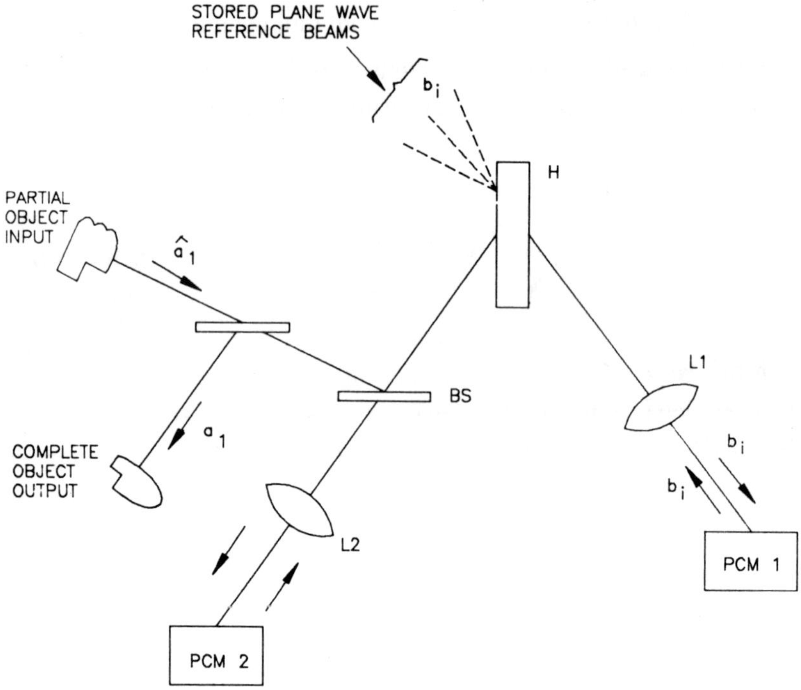

Figure 14.8 Schematic arrangement for nonlinear optical associative memory proposed by Soffer *et al* (adapted from Ref 43)

wavefront used to record the *i*th hologram. Suppose now we want to reconstruct the wavefront a_i from a wavefront $\hat{a}_l$ which carries part of the image depicted by a_l. The arrangement proposed by Soffer *et al* (1986) to do this is shown in Figure 14.8. H is the composite hologram mentioned above that contains the interference patterns of wavefronts $\{a_i\}$ with corresponding plane reference waves $\{b_i\}$, PCM_1 is a thresholding phase conjugate mirror whose reflectivity depends upon the input irradiance. This could be a PCM based on self-pumped DFWM (Ridley and Scott 1990) or on SBS. The second conjugator PCM_2 is an externally pumped phase conjugation mirror whose reflectivity is independent of the input irradiance. The partial wavefront $\hat{a}_l$ on diffraction by hologram H produces the corresponding plane wave b_l with some distortion. On being focussed (Fourier-transformed, neglecting the phase factor) by lens L_1 the plane wave component of the diffracted beam produces a higher irradiance image. Because of the irradiance-dependent reflectivity of PCM_1 this component is preferentially generated in the phase-conjugate reflected beam. On being essentially Fourier-transformed by lens L_1 this produces a relatively less distorted plane wave b_l. This when diffracted by H produces a better approximation of the complete wavefront a_l which on being reflected by PCM_2 will start the next cycle.

Obviously there will be diffraction losses in each cycle as well as some other losses. Thus the complete object wavefront will be generated without distortion after many cycles provided the gain in PCM_1 and PCM_2 is sufficient to overcome these losses. We should also note that, by the very nature of associative memories, if the partial information has substantial overlap with several images the arrangement could produce any one of them or may not converge to any one.

There are many other applications of DFWM in optical computing. Of these schemes those in which the polarization of the beam is used to carry the information are of particular interest because of their easy cascadibility. For more details reference may be made to some recent papers (O'Meara 1982, Kumar *et al* 1990).

14.7 Conclusion

We have given a brief introduction to the physics of optical phase conjugation and degenerate four-wave mixing which is probably the most important method of obtaining optical phase conjugated beams. Some applications of OPC and DFWM have been described mainly with the intention of illustrating the new concepts involved and not necessarily because they are close to commercial exploitation. Many important aspects could not be included. Some of these are the investigations of ultrafast phenomena using DFWM, quantum optical aspects such as squeezed states, many important applications of OPC/DFWM using photorefractive crystals, phase conjugating resonators and some very exciting applications such as correction of modal dispersion in multimode fibres and the phase conjugate fibrescope. It is hoped that in spite of its shortcomings this overview would help arouse the readers' interest in this fascinating field.

Acknowledgements It is a pleasure to thank Mr Selvakumar Nair for help in the preparation of the manuscript and, more importantly, by way of critical readings of various parts of it.

References

CARO R G and Gower M C, 1982 IEEE J Quant Electron **QE-18** 1376

COTTER D, Electronics Lett 1986 **22** 693

DUTTON T, Dvornikov A, Van Wonterghem B and Rentzepis P M, 1990 Appl Phys Lett **57** 1952

EICHLER H J, Gunter P and Pohl D W, 1986 *Laser Induced Dynamic Gratings* (Berlin: Springer-Verlag) and references therein

EICHMANN G, Li Y, Ho P P and Alfano R R, 1987 Appl Opt **26** 2726

EWART P and O'Leary S V, 1986 Opt Lett **11** 279

FISHER R A, 1982a Ed *Optical Phase Conjugation* (New York: Academic Press)

FISHER R A, 1982b *Optical Phase Conjugation* in Encyclopaedia of Physical Science and Technology Vol 9 Ed Meyers R A (New York: Academic Press)

GIULIANO C R, 1981 (April), Physics Today p 1

GOWER M C, 1985 *Progress in Quantum Electronics* Vol 9

HACHE F, Roussignol P, Ricard D and Flytzanis C, 1987 Opt Comm **64** 200

HELLWARTH R W, 1977 J Opt Soc Am **67** 1

HORAN P and Blau W, 1990 *Phase Transitions* **24-26** 605

ISHIOKA Y and Tanida J, 1989 Optoelectronics - Devices and Technologies **4** 39 and references therein

JAIN R K and Klein M B, 1982 in Fisher 1982a

JAIN R K and Lind R C, 1983 J Opt Soc Am **73** 647 (Special issue on Optical Phase Conjugation)

LEVENSON M D, 1983 J Appl Phys **54** 4305 and references therein

LEVENSON M D and Chiang K, 1982 IBM J Res Dev **26** 160

MAXWELL-GARNETT J C, 1904 Philos Trans R Soc Lond **203** 395

NEFF J A, Athale R A and Lee S H, 1990 Proc IEEE **78** 826

NOSACH O Y, Popovichev V I, Ragul'skii V V and Faisullov F S 1972 JETP Letters **16** 435

NUSS M C, Zinth W and Kaiser W, 1986 Appl Phys Lett **49** 1717

O'MEARA T R, Pepper D M and White J O, 1982 in Fisher 1982a (*loc cit*)

OWECHKO Y, 1989 IEEE J Quant Electron **QE-25** 619

PARK S H, Morgan R A, Hu Y Z, Lindberg M, Koch S W and Peyghambrian N, 1990 J Opt Soc Am **B7** 2097

PEPPER D M, 1985 in *Laser Handbook* vol 4 Ed Stitch M L and Bass M (Amsterdam: North-Holland)

PEPPER D M, 1986 (January), Sci Am **254** 74

RAKESTRAW D, Farrow R L and Dreier T, 1990 Opt Lett **15** 709

RAVINDRA KUMAR G, Singh B P, Rao K D and Sharma K K, 1990 Opt Lett **15** 245

RIDLEY K D and Scott A M, 1990 Opt Comm **76** 406

ROUSSIGNOL P, Ricard D and Flytzanis C, 1990 Appl Physics **B51** 437

ROUSSIGNOL P, Ricard D, Rustagi K C and Flytzanis C, 1985 Opt Comm **55** 143

RUSTAGI K C and Bhawalkar D D, 1990 Ferroelectrics **102** 367

RUSTAGI K C and Flytzanis C, 1984 Opt Lett **9** 344

SHEN Y R, 1984 *Principles of Nonlinear Optics* (New York: Wiley-Interscience)

SHKUNOV V V and Zel'dovich B Ya, 1985 (December) Sci Am

SOFFER B H, Dunning G J, Owechko Y and Marom E, 1986 Opt Lett **11** 118

TAKAGAHARA T, Phys Rev **B36** 9293

TOMITA T, Yahalon R, Kyúma K, Yariv A and Kwong N S K, 1989 IEEE J Quant Electron **QE-25** 315 (Special issue on Optical Phase Conjugation)

WEBER C, Becker U, Renner R and Klingshirn, 1988 Z Phys B Cond Matter **72** 379

WOERDMAN J P, 1970 Opt Comm **2** 212

WOERDMAN J P and Bolger B, 1969 Phys Lett **32A** 164

WOLF E, Agarwal G S and Frieberg A T, 1983 in Proc 5th Rochester Conf on *Coherence and Quantum Optics* Ed Mandel L and Wolf E (New York: Plenum)

YAO Li, Eichmann G, Dorisinville R and Alfano R R, 1988 Appl Opt **27** 2025

YARIV A, 1985 *Optical Electronics* 3rd ed (New York: Holt, Rinehart and Winston)

YATAGAI T, 1986 Opt Lett **11** 260

ZEL'DOVICH B Ya, Pilipetskii N F and Shkunov V V, 1985 *Principles of Phase Conjugation* (Berlin: Springer-Verlag)

ZEL'DOVICH B Ya, Popovichev V I, Ragul'skii V V and Faisullov F S, 1972 JETP Letters **15** 109